Global Rhetorics of Science

SUNY series, Studies in Technical Communication
———
Miles A. Kimball, Charles H. Sides, Derek G. Ross,
and Hilary A. Sarat-St. Peter, editors

Global Rhetorics of Science

Edited by
LYNDA C. OLMAN

Cover credit: Anne Dirkse, *Moai Under the Milky Way at Ahu Tongariki, Easter Island*, Creative Commons Attribution-Share Alike 4.0, https://creativecommons.org/licenses/by-sa/4.0/.

© 2023 State University of New York, Albany

All rights reserved

Printed in the United States of America

No part of this book may be used or reproduced in any manner whatsoever without written permission. No part of this book may be stored in a retrieval system or transmitted in any form or by any means including electronic, electrostatic, magnetic tape, mechanical, photocopying, recording, or otherwise without the prior permission in writing of the publisher.

For information, contact State University of New York Press, Albany, NY
www.sunypress.edu

Library of Congress Cataloging-in-Publication Data

Name: Olman, Lynda C. (Lynda Christine), 1971– editor.
Title: Global rhetorics of science / edited by Lynda C. Olman.
Description: Albany, NY : State University of New York Press, [2023] | Series: SUNY series, studies in technical communication | Includes bibliographical references and index.
Identifiers: LCCN 2022053654 | ISBN 9781438494432 (hardcover : alk. paper) | ISBN 9781438494449 (ebook) | ISBN 9781438494425 (pbk. : alk. paper)
Subjects: LCSH: Communication in science. | Science—Language. | Rhetoric.
Classification: LCC Q223 .G563 2023 | DDC 808.06/65—dc23/eng/20230608
LC record available at https://lccn.loc.gov/2022053654

10 9 8 7 6 5 4 3 2 1

Contents

List of Illustrations	vii
Preface In Memorium: Ubiratan D'Ambrosio, Ethnomathematics, and Rhetoric	ix
Acknowledgments	xiii
Introduction: Reconfiguring Global Rhetorics of Science Lynda C. Olman	1
Chapter 1 How Euro-American Science Became Dominant: Transnational Circulations of Knowledge and Capital Kelly Happe and Lynda C. Olman	23

Chapter 2
The Shifting Rhetoric of Environmental Science in Australia: Acknowledging First Nations People and Country
    Emilie Ens, Shaina Russell, Bridget Campbell,
    Sabina Rysnik-Steck, Monica Fahey, Patrick Cooke,
    Renee Cawthorne, and Daniel Sloane
	45

Chapter 3 African Sciences and Indigenous Knowledge Systems in the West African Ebola Crisis Toluwani Oloke and Olusegun Soetan	63

Chapter 4
A Critical Contextualized Approach to Studying Clashing Risk
Cultures: Mapping the Transcultural Environmental Risk
Communication of PM2.5 in China   87
   Huiling Ding and Jianfen Chen

Chapter 5
Where Voyaging Ends: Social Cosmology on Rapa Nui   113
   Francisco Nahoe

Chapter 6
Celtic Geometric Art as a Visual Rhetoric of Science   139
   Evelyn Dsouza

Chapter 7
This Is a Viral Story about Viral Stories: Image and Graphical
Power in COVID Communication in the Navajo Nation   157
   *Sunnie R. Clahchischiligi, Julianne Newmark,
   and Joseph Bartolotta*

Chapter 8
A Rhetoric of the Home Ground: Local Knowledge and
Data-Gathering among the North Atlantic Glaciers   179
   Ryan Eichberger

Bibliography   203

Contributor Biographies   233

Index   239

Illustrations

Figures

3.1	Participants at the Ebola awareness training organized by ASEOWA in Grand Cape Mount, Liberia.	78
3.2	Town Hall Meeting 1.	80
3.3	Town Hall Meeting 2.	81
3.4	African Union ASEOWA joint awareness collaboration with Liberian Police Force.	81
4.1	AQI levels and meanings released by the US Embassy in Beijing.	98
4.2	Cartoon: "I Gauge the Air Quality for My Motherland."	100
4.3	Screenshot of Pan's Weibo Post on October 22, 2011.	102
5.1	The moai quarry at Rano Raraku.	114
5.2	Ahu a Kivi, restored in 1960 by William Mulloy and Gonzalo Figueroa.	120
5.3	The ceremonial complex at Tahai, restored by William Mulloy in 1974.	124
5.4	Ahu Vai Ure at Tahai with the outskirts of the modern village of Haŋa Roa in the background.	127
6.1	An example of rotational symmetry: a decorative knot with 15 crossings.	000

6.2	A triskele, based on motifs found at megalithic tombs in Newgrange, Ireland.	143
6.3	A zoomorphic bronze-cast Celtic brooch found in southwestern Germany.	144
6.4	A triquetra.	147
7.1	"Keep Ur Tl'aa' Home."	158
7.2	COVID-19 graffiti.	165
7.3	"Stay Home."	170
7.4	"Be the Social Vaccine."	173

Table

4.1	Timeline for risk communication of PM2.5 and hybrid argument.	95

Preface in Memorium

Ubiratan D'Ambrosio, Ethnomathematics, and Rhetoric

> Metaphorically, I consider disciplines as cages: it is not possible to leave the cage, since wires impede it. The wires are codes. I consider codes in the broad sense including symbols, jargon, criteria of truth, rigor and precision and other normative specificities. In the cage, it is not possible even to know the color of the external paint! The search for knowledge within the cage reveals nothing about what is outside.
>
> —Ubiratan D'Ambrosio, "Rhetoric and Ethnomathematics"

It was with great sadness that the authors of this volume learned, in July 2021, that one of our number had passed away a few months earlier in May. When we lost Ubi, we lost not only the premier ethnomathematician in the world, the founder of the discipline, but also a tireless champion of this volume and the decolonizing effort it represents. When the pandemic was at its height, and it looked like we might not be able to complete the project, Ubi wrote to me, "Don't worry. Nothing will stop this beautiful book from being published." His encouragement carried all of us forward.

Ubiratan D'Ambrosio was Professor Emeritus of Mathematics at the State University of Campinas (UNICAMP) in São Paulo. Born in 1932, a lifelong citizen of Brazil, he achieved his PhD in mathematics at the young age of twenty-one. Throughout his lengthy and illustrious career, Ubi held academic positions in Italy, the United States, and Brazil. He was a Fellow of the American Association for the Advancement of Science (AAAS) and the member of multiple illustrious international STEM education commissions. He also co-organized the Nobel Prize–winning Pugwash Conferences on Science and World Affairs.

Ubi was generally acknowledged as the founder of the discipline of ethnomathematics, which he defined in his draft chapter for this volume as "the specific corpus of knowledge dealing with quantitative and qualitative practices, such as counting, weighing and measuring, comparing, sorting and classifying, and inferring, accumulated through generations in distinct natural and cultural environments." Since its inauguration in 1977, the field has concentrated both on recovering indigenous mathematical practices and on critiquing the ways in which Euro-American mathematics,[1] and mathematics education, have been implicated in colonization. For his part in sustaining these programs of research—not only through his leadership but by publishing over 300 academic essays in 6 languages on ethnomathematics and education—Ubi was awarded the 2001 Kenneth O. May Medal from the International Commission on the History of Mathematics (ICHM) as well as the 2005 Felix Klein Medal from the International Commission on Mathematical Instruction (ICME).

In the preliminary draft of "Rhetoric and Ethnomathematics" that he prepared for this volume, Ubi took a deeply philosophical approach to the contact between European and indigenous mathematical systems in the colonial invasion of the Americas. He did so via Program Ethnomathematics (PE), which treats encounters between knowledge systems from a transdisciplinary and rhetorical perspective. Ubi argued that disciplines are supported and expressed via codes of symbolic communication, including language. He revisited the colonization of Brazil in the 16th century to show how codes of Euro-American mathematics influenced the quantification of everything from plants to people, and how they restricted the ability of the European colonizers to recognize and respect the epistemologies of the indigenous peoples they encountered. Ubi wrote, "Science and mathematics were present in a very sophisticated way, epistemologically different from European knowledge. So radical were these differences for Amazonian cultures that the colonizers concluded they lacked mathematical concepts altogether." This failure of recognition served a rhetorical function for the colonizers; it "supported efforts to label the encountered cultures as primitive, uncivilized and justified the treatment given by Europeans to the indigenous peoples." He explained that "the main interest of the conquerors was to explore the native knowledge about extraction of minerals, particularly gold and precious stones, the use of plants and animals. It was important to dissociate these products from any . . . native right; thus, the importance of associating them with Western interests via a sort of elementary scientific indoctrination conducted by catechists." Ubi devised PE as a direct response to this colonial erasure of indigenous epistemologies. PE operates primarily through transdisciplinarity, or "the abolishment of all cages." It doesn't seek to erase Euro-American mathematics; such a goal would be impossible. However, PE remains radically open to the hybridization of Euro-American and indigenous mathematics, with the ultimate goal being the development of a common rhetoric. Ubi gave the example of work by Daniel Everett and others who found that, when Amazonian tribes work math problems, they use Euro-American formulas but perform "the operations from bottom to top, explaining that this is the way trees grow." Instead of correcting these practices as faulty, PE frames an educational program that can "embrace new characteristics of rigor in the discourse; of precision in time and space; and open language (jargon), codes, etiquette and social stratification. This revision requires openness, coherence and respect."

Ubi's call for openness, coherence, and respect is in many ways the rallying cry for this volume. Like Ubi, we do not seek to eradicate Euro-American STEM practices in post- and neo-colonial contexts, but rather to negotiate a new epistemological and rhetorical contract that recognizes indigenous sciences and makes them equal partners in the management of global risks. We are proud and grateful to have had Ubi point us the way to this goal, and to encourage our first steps.

Note

1. On the choice of *Euro-American* in place of *Western* to describe the current dominant global scientific tradition, please see the introduction to this volume.

Acknowledgments

Thanks to our series editors, Miles Kimball, Charles Sides, Derek Ross, and Hilary Sarat-St. Peter, for shepherding this project through a difficult epoch in the field and in the world. We're grateful as well to our editors at SUNY Press: James Peltz, Tim Stookesberry, and Richard Carlin. Thanks are also due to the three anonymous reviewers of this project whose time and care resulted in a much better volume than would have been published otherwise. Finally, we honor the Indigenous peoples whose ancestral lands hosted our work and the institutions that employed us during the composition of this volume; these groups, who will be named in the chapters to follow, include Australian Aboriginal, Pacific Islander, Central Asian, Native North American, West African, Arctic, and Celtic European peoples.

Additionally, Francisco Nahoe thanks Sonia Haoa, Phineas Kelly, Gerald Nahoe, and Josefina Nahoe, all of whom read early versions of the chapter and offered critiques that helped him to improve his account of the *tupuna*, and readily acknowledges that any remaining defects belong to him alone.

Ryan Eichberger wishes to thank the people whose words and time became the substance of his chapter: in order of appearance, Michael Kienitz, Snævarr Guðmundsson, Þorvarður Árnason, and Kerstin Langenberger. Thanks are due also to Cymene Howe and Dominic Boyer, who generously shared their thoughts about glaciers and film. Additionally, thanks to those who connected Ryan with materials at libraries and collections around the world during a pandemic: Sigurður Stefán Jónsson, Oddur Sigurðsson, Patrick Joseph Stevens, and Richard S. Williams Jr. And thanks to Abraham Swee for his support through long nights of typing and his forever-willingness to hike up and down cliffs and coasts. Finally,

Ryan wishes to thank Daniel J. Philippon, who gave feedback throughout the first versions of his chapter, and who encouraged him to write what seemed important. He is indebted to them all.

Introduction

Reconfiguring Global Rhetorics of Science

Lynda C. Olman

> Is there an equivalent in traditional knowledge to what science calls a theory? Absolutely. But it's a different kind of theory, one that centers on the idea of responsibilities. All bees, for example, have a responsibility to pollinate. The Indigenous observer is asking the bee, How are you living out your responsibility? And what about you, flower?
>
> —Robin Wall Kimmerer[1]

> We cannot solve our problems with the same level of thinking that created them.
>
> —apocryphal, attributed to Albert Einstein

A story told in *Desert Lake: Art, Science, and Stories from Paruku* explains the purpose of this volume better than I can. *Desert Lake* recounts a joint scientific-artistic project involving the Walmajarri, traditional custodians of Paruku (Lake Gregory) in Western Australia. The Paruku Project began when the Walmajarri asked for assistance in controlling a fish parasite, as well as in assessing several archaeological sites, including one that helped pushed back current knowledge of the presence of humans in Australia to around 50,000 BCE. For their part, the collaborating Euro-American scientists[2] received unprecedented access to these sites and new data on

Aboriginal ecological management practices. The project was not without its complications, however, among these an incident in which custodians brought archaeologists to examine some bones that had been unearthed by rains. Community lore framed these as the remains of a murder or ancient conflict. But the archaeologists quickly identified a proffered tooth as nonhuman, likely from a horse. Notwithstanding, one custodian held up the tooth before the community and declared it human. The archaeologists strove to correct the record, speaking over the custodian, at which point an elder put her hand on one of the scientists' arms gently and said, "You're not listening." The archaeologists subsided, leaving the tooth to hover in collective awareness in superposition—both horse and human.[3]

It was both an emblematic and a pivotal moment for the Paruku Project: emblematic in that it epitomized the many, many incommensurabilities between globalized Euro-American and Indigenous knowledge-ways that had doomed previous attempts at collaboration with the Walmajarri;[4] pivotal in that the Project was able to move past it by layering rather than reducing. Instead of insisting that the human tooth be replaced by the horse tooth in viewers' minds—in other words, instead of insisting that Indigenous ways of knowing and representing the world must either agree with or be reduced to Euro-American ways—the project's organizers chose instead to let both interpretations stand, equally true and valuable, each meeting a particular need for the people who called it forth. The organizers accomplished this layering via several techniques, chief among them the literal layering of acrylic paints over topographic maps of key sites, like the tooth dig site, to create paintings of Country (land/history/culture). The ultimate success of the project graphically made the point that global risks—including climate change, pandemics, and food and energy security—cannot be collectively and effectively managed if we continue to insist that all global systems for knowing and representing natural phenomena be reducible to the Euro-American scientific system. That is a point the present volume takes both very seriously and as a starting point from which to search for new ways to integrate global sciences in the just and collective management of global risks.

Most of the content of this starting point—the part about the equality of human systems of natural knowledge and the effacement of that equality by colonial dynamics—is old news, established amply by scholarship in science and technology studies (STS) and the history of science over the last fifty years. While a review of this work is beyond the scope of this introduction (though interested readers will find a list of key texts in the notes), it has thoroughly dispelled the myth of science and scientific methods as Euro-American inventions.[5] This scholarship has convincingly demonstrated the dependence of Euro-American science on colonization as well as its emergence from older global traditions, most notably the Islamic sciences, from which Euro-American science took (and took credit for) foundational concepts and theories in mathematics, astronomy, geography, chemistry, and medicine. That old saying about the only difference between a dialect and a language being that the latter has an army and a navy applies equally to the difference between "science" or "medicine" on the one hand and "ethnoscience" and "alternative medicine" on the other: in other words, the dominance of Euro-American science today in the world is more the result of colonial geopolitics than of any difference in quality among global sciences, or any process of systematic verification.

The need for collective management of global risks across traditional boundaries and borders is a relatively new concern, however. Evidence is mounting that top-down, neoliberal approaches to managing these risks produce significantly negative consequences for vulnerable communities, particularly in the Global South.[6] What is needed instead are risk collectives that can nimbly integrate global and local information to manage global risks equitably for these communities.[7]

The good news here is that the collective management of risk and uncertainty has been the business of rhetoric for the last 2,500 years in the Euro-American tradition and much longer in other global traditions of communication like the Aboriginal practices of Country and Law. Furthermore, the move to reduce all global sciences to Euro-American science is a rhetorical move, an act of synecdoche in which a part of a system is made to stand in for the whole. Accordingly, we are organizing this volume around rhetorical tactics for moving away from synecdoche and into other ways of configuring the relationship of Euro-American science to its fellow sciences around the globe. In other words, if the problem is that we've been treating Euro-American science as a synecdoche for all global sciences, then we can start to reinvent that relationship by studying cases of subversion and resistance to the hegemony of Euro-American science (irony); imagining new rhetorics of science by taking inspiration from non-Euro-American ones (metaphor); and writing narratives about the world that collect and associate rather than reduce and analyze (metonymy). While this reconfiguration is admittedly a very limited decolonization of the rhetoric of science (ROS), it has the advantage of working within our existing toolkit to start shifting the discourse.

While the authors in this volume will name and describe the global communication traditions in which they are working, it is my task to provide a brief orientation to ROS. In the spirit of the volume, I wish to view this history with a somewhat wider aperture than is customary, taking in the coevolution of ROS with Islamic rhetoric and science—in acknowledgment of the heavy debt that the Euro-American traditions owe in this regard.[8]

In the Euro-American tradition, rhetoric was invented by a collective of Attic Greek tribes in the 6th century BCE as a social technology used to get the collective to act together for the purposes of managing the uncertainties they jointly faced—both environmental and social. It was officially organized and codified as *techne*, a technology or productive art, in the 4th century BCE by the Socratics, Aristotle chief among them. He famously defined rhetoric as "an ability, in each particular case, to see the available means of persuasion."[9] It was during this period that the first serious contact and hybridization between Greco-Roman and Arabic traditions occurred, as the library at Alexandria became an important node for intercultural exchange of scholarship and art. Greek science became influenced by the classical Babylonian sciences—particularly astronomy, agronomy, and medicine—developed by eastern Sassanid (Iranian/Persian) and Hindu scholars and syncretized by the Arabic scholars working at Alexandria. At the same time, these scholars translated and studied Greco-Roman rhetorical and scientific theories and integrated them with their own.

Meanwhile, in Rome, the art of rhetoric was being developed and fine-tuned by senators and educators such as Cicero and Quintilian. As empire gradually eroded democracy, however, rhetoric's roots in the collective management of risk also withered, and the art became transplanted largely into scholastic grounds, where it survived through the political chaos of the Middle Ages, cloistered in monasteries and universities. During this same time frame, scholarship in the sciences and communicative arts was actively advancing under the Umayyad (661–750 CE) and Abbasid (750–1258 CE) caliphates. A key factor in the explosion of Islamic learning during this Golden Age was a view of Allah as a creator who not only permitted but actively promoted the investigation of his creation as a form of devotion.[10] This principle enabled an interlinkage of religious, artistic, and scientific activity that caused science to flourish in Islam, while Aristotle's texts were quarantined in Byzantine treasuries as dangerous pagan heresy.[11] It was through the raiding of these treasuries in the Crusades, alongside other territorial skirmishes throughout the

Mediterranean, that the second great rapprochement of European and Islamic traditions came about.

Historians of science used to restrict the contributions of Islamic scholars to the development of European science to the mere safekeeping and translation of Greco-Roman scientific texts during Europe's dark ages, with grudging allowances made for the innovations of algebra and alchemical experimentation. However, this story has been revised over the last forty years as science historians began to read Arabic texts in the original, and as historians of Islam became interested in the origins and dissemination of scientific knowledge. We now know that the books that Thomas Aquinas and other European scholars gleaned from the bloodstained conflicts in the Levant were not mere translations of Greek and Roman scholarship but significant innovations on it—many of which the Europeans simply assimilated without crediting the innovators. For instance, William Harvey's theory of blood circulation was anticipated by Ibn an-Nafis's four centuries earlier; the revision of planetary motion attributed to Copernicus was substantially worked out by Ibn al-Shātir in the 14th century; and systems of geographic projection worked out by Arab geographers in the 9th to 12th centuries enabled a quantum leap forward in European mapmaking and navigation in the 16th century.[12] While work is ongoing to prove direct lines of influence in some of these cases, we do know that original texts by Islamic scientists were circulating in Europe well in advance of the Renaissance and almost certainly played a significant role in its emergence.[13] The Islamic tradition emphasized the practical and political applications of scientific knowledge well before this focus emerged in 17th-century Europe.[14]

After this second period of cross-fertilization, geopolitical circumstances on both sides—namely, imperial colonialism in Europe and the conservatism of the Ottoman Empire—effectively cut off collaboration between the respective scholarly communities. This development made it hard to challenge the growing myth of an autochthonous European science, and eventually that myth became dominant, acknowledging Euro-American science as the only valid science. However, in the last fifty years, as a result of globalization and a new wave of geopolitical conflicts in the Middle East, we have experienced a third engagement of Islamic and Euro-American knowledge systems that has helped challenge the Euro-American hegemony.

As Islamic nations around the world have gained population, wealth, and political clout, some of their scholars have called for a return to an

Indigenous Islamic science, under the heading of Islamization of Knowledge (IOK), as a means of decolonizing their societies. Turkey recently banned the teaching of evolution in public schools,[15] and Islamic schools teaching IOK-based curricula are proliferating in Malaysia, South Africa, and Europe, among other places.[16] Scholars studying the IOK movement are deeply divided over its politics: some see in it the promises of innovation, autonomy, and cultural healing, while others fear that only balkanization and fundamentalism will result.[17] But the IOK debate stands as the first major attempt to engage questions very similar to those engaged by this volume, and to recognize these questions as not merely ones of knowing the world, but making it: How do we move past the neocolonial era into a more equitable global cooperation in technoscience and politics? What will this new era look like? The Islamic principle of the divine unity of all human endeavor is one source of the promise that scholars such as Osman Bakar locate in Islamic sciences—to point a path forward out of our global risk crises.[18]

A related move toward the unification of world-building (science) and community-building (rhetoric) has occurred in rhetorical studies over this same epoch. Though reduced on the European continent to the study of stylistics by the 19th century, the preaching tradition in England kept Classical principles of oratory alive and functioning there; with the advent of the Belles Lettres era, rhetorical techniques began to be used not just as heuristics for the creation of persuasive texts, but also as hermeneutics for their interpretation. This is the tradition that became regularized as current traditional pedagogy in land-grant universities in the United States, which had a mission to educate all post-secondary students up to an acceptable professional standard of reading and writing.

After World War II, the neocolonial globalization of Euro-American capitalist democracy—and its attendant risks—prompted a rebirth of rhetoric not just as composition pedagogy but as social *techne*. Scholars such as Kenneth Burke, Chaim Perelman, and Lucie Olbrechts-Tyteca returned the focus of the art to collective deliberation;[19] meanwhile, sociologists such as Ulrich Beck were using terms like *world risk* to describe the new exigence for collective action in late capitalism.[20] In 1982, Thomas Goodnight coined a new definition of rhetoric for the era: "The creative resolution and resolute creation of uncertainty."[21] Since that time, scholars of rhetoric have turned to excavating the ancient roots of rhetoric in collective risk management, focusing in particular on the rhizomatic action of rhetoric as a middle ground between critical paralysis on the one hand, the totalizing solutions to risk management on the other,[22] and the circulation of nonsymbolic material and energy in the flow of communication.[23]

This excavation effort naturally led some rhetoricians outside the boundaries of the Greco-Roman tradition to look for alternatives to its agonistic, imperialist tendencies. This new field of comparative rhetoric (originally contrastive rhetoric) was reflexive from the beginning, pointing out the dangers of hunting for Euro-American concepts in non-Euro-American settings even as it sought an ethical footing from which to conduct crucial discovery and recovery work.[24]

This debate was soon echoed in decolonial studies of rhetoric, which have over the course of the last decade aimed not only at criticizing the hegemony of Euro-American rhetorics in the "contact zones" (primarily American ones) generated by colonial imperialism, but also at counteracting and even undoing imperialism's effects.[25] For example, in her study of Sarah Winnemucca and Charles Alexander Eastman's oratorical practices, Malea Powell cautions against the merely "additive" approach to decolonizing rhetoric by expanding the canon, arguing that this still centers and privileges Euro-American rhetoric.[26] Along these same lines, Ellen Cushman has argued that, in a truly decolonial framework for rhetorical education, the dominance of the Euro-American tradition—particularly, the English language's "place as a lingua franca"—must be "questioned, equalized, and replaced with a more cosmopolitan understanding of English's place alongside and equal to a pluriversality of languages."[27] But in order to accomplish this decentering, the "discipline's tendency to prioritize so-called objective approaches to knowledge and Euro-American narratives of rhetorical practice" must first be challenged, according to Lisa King and the editors of *Survivance, Sovereignty, and Story: Teaching American Indian Rhetorics*.[28] The pivotal role of objectivity in the dominance of Euro-American disciplinary traditions is the point at which ROS must become involved in the decolonization project.

Unfortunately, ROS has lagged behind in comparative and decolonial work. ROS began in the 1970s and 1980s by studying how Euro-American scientists used words—in addition to equations and experiments—to persuade members of their own professional communities to accept their claims about the natural world. With the critical turn in the 1990s, ROS widened its view to consider the interaction of scientific communities with the societies supporting them.[29] Since then, the field has made great strides in diversifying "rhetoric" into "rhetorics," considering the impact of economic class, race, gender, disability, and nonhumans on the way science is done and communicated.[30] But rhetoric's partner term, "science," has remained stubbornly monolithic—so much so that, to this day, if readers in STS restricted themselves only to ROS scholarship they would never know other global sciences existed. Meanwhile, as noted above, historians, philosophers, and sociologists of science have been investigating global sciences for 200 years, intensively so for the last 50.[31] Even in ROS's sister field of technical and professional communication (TPC), scholars have made much more progress in decolonizing the discipline, for instance, by recuperating the professional contributions of people of color to science and technology,[32] critiquing the colonial agency of TPC in vulnerable communities,[33] revising fieldwork methodologies for studies in post- and neocolonial settings,[34] and decolonizing TPC pedagogy.[35]

There are several reasons for ROS's failure to interrogate the dominance of Euro-American science, such as a laudable preoccupation with watchdogging its abuses in vulnerable populations, as well as an understandable reluctance to jeopardize collaborations with scientists by appearing "anti-science." There is also, perhaps, a less-admirable obsession with the core Euro-American canon of STS, born out of a desire to gain admittance to the club, so to speak. But the reasons hardly matter at this point. What matters is what has always mattered to rhetoric: *now*. And *now*, science—more accurately, the "triple helix" of state, corporate, and university actors that together constitute the sociopolitical agency of Euro-American science[36]—has materially participated in bringing about the global environmental crises we face; *now*, our colleagues in STS and history of science are decades ahead of us in decolonizing scientific practice. It is well past time for rhetoricians to join the fray—particularly because what is needed right *now* is expertise in deliberating across differences to collectively manage global risk, and that is our wheelhouse.

Not only will a shift from "rhetorics of science" to "global rhetorics of science" generate new, useful strategies for collective risk management, it will also give ROS a mirror in which it may reflect on its own practices, including its continued privileging of Euro-American rhetorics. And it will hopefully contribute to a rapprochement between Euro-American and global lifeways at a time when mutual balkanization and fundamentalism are creating situations like the current one in Turkey or the recent one in the United States, in which the president hamstrung his own Environmental Protection Agency and repeatedly and publicly denied the reality of global climate change.[37]

In sum, rhetoricians not only have much to gain from studying global rhetorics of science; we also stand to lose a great deal if we continue to give our tacit support to the imperialism of Euro-American science, particularly when it fits hand in glove with neoliberal, transnational interventions in the lives of Black women in the United States, coastal fisherman in Indonesia, and farmers in Nicaragua.[38] If as rhetoricians we remain committed to democracy and justice, it stands to reason we should be helping to reinvent these neocolonial dynamics.

But how to begin? Certainly, one volume cannot and should not hope to achieve the decolonization of ROS. Accordingly, as editor, I set a modest goal for our project: to use what was already in our toolkit as rhetoricians to take a step outside our Euro-American nursery—both to look back at it and to look outward, to *reconfigure* the relationship between Euro-American science and global sciences. I have structured this reconfiguration using the four most common or dominant rhetorical figures, formerly referred to as "master tropes": synecdoche, irony, metaphor, and metonymy. Historian Hayden White has argued persuasively that these tropes frame the majority of modern histories,[39] so they seemed like a good starting place for resetting Euro-American science in its proper context as one among many global sciences. Plus, I relished the idea of *turning* (troping) this colonial frame of "master" tropes back on itself in a decolonial project. Troping, as a mainstay of colonial rhetorics, has also served as a key site for decolonial intervention.[40] And so it will serve in the present volume.

In what follows I will briefly define the four dominant tropes with examples and explain how they frame the contributions to this volume before concluding this introduction.

Synecdoche: Reducing a Complex Whole to One Part

Synecdoche is a rhetorical figure of reduction. It treats a complex idea by way of one part that is easier to grasp and manipulate.[41] So, for instance, in the United States, we say "Washington is in chaos" when concerned about overall disarray in our federal government, or we use the synecdoche "main street" to encapsulate the attitudes of millions of Americans who live in small cities and towns. President Obama was an expert in synecdoche, framing his speeches around the lived experience of this grandmother or that teenager, their individual life expressly chosen to stand in for the lives of their wider communities, perhaps even the entire nation.

Because it serves to reduce massive, complex situations to small ones that are easier to comprehend and control, synecdoche confers a great deal of rhetorical power on its user. Unsurprisingly, then, synecdoche has played a central role in the global dominance of Euro-American science, as this singular scientific tradition came to stand in for *all* global scientific traditions in education, health care, and environmental policy-making. Now, if Native American children want to become scientists, they must be inducted into a scientific tradition born an ocean away from their homeland and, in the process, repress or reject much of their own Indigenous knowledge of their world.[42] By the same token, when autochthonous accounts of natural phenomena conflict with the globalized Euro-American account, they must be rejected or reduced to fit. This Procrustean reconciliation happens constantly, but some recent examples include the tragic governmental dismissal of local warnings against the L'Aquila earthquake in Italy;[43] the destruction of traditionally managed mangrove lagoons to build costly, ineffective, and unsustainable seawalls in Indonesia;[44] and the decades-long refusal by the United States Food and Drug Administration (USFDA) to approve drugs to treat fibromyalgia since it was considered a hysterical condition existing only in "crazy" women's heads.[45]

Naturally, these kinds of rhetorical reductions would never have succeeded if they had not been wedded to a major geopolitical reduction—namely, neoliberal capitalistic globalization. Accordingly, Kelly Happe dedicates the first chapter of this volume, "How Euro-American Science Became Dominant: Transnational Circulations of Knowledge and Capital" to tracing the history of this geopolitical reduction, using reproductive technologies as a case study to illustrate how something as rooted, local, and individual as a human ovum came to be the globalized object of Euro-American scientific definition, control, and capitalization. The chapter describes and unpacks the ways in which capitalist logics inform the "the capitalization of life" and the "co-production" of science and market logics when ova become a kind of speculative biocapital. Happe finishes by looking forward—suggesting ways to reconfigure the synecdochal reduction of life and thereby decapitalize it.

Irony: Exposing Diversity in Apparent Unity

Most readers are probably familiar with irony from their high-school lessons about one particular species of it—dramatic irony, in which the assumptions of the protagonists are shown to be unfounded by developments that reverse them. But rhetorical irony is a more general and capacious figure that reveals discontinuity and dissensus in situations assumed to be continuous and consensual.[46] So, when Mark Antony repeats "Brutus is an honorable man" in his funeral oration, he uses that phrase as an ironic lever to pry open the consensus on Brutus's honor. Satire unsurprisingly trades heavily in irony, as did Jonathan Swift's "Modest Proposal" in the 18th century that the English should simply eat excess Irish children to control their population and spare them the torments of poverty.[47]

As is clear in the Swift example, irony is also the organizing trope of critique. And in reconfiguring the relationship between Euro-American science and other global sciences, critique is an important first step—to render visible currently invisible hegemonic colonial (or neocolonial) power dynamics. Three of our chapters contribute to the decolonization of ROS by critiquing the rhetorical and political power wielded by Euro-American science in global—particularly colonial—contexts. In chapter 2, "The Shifting Rhetoric of Environmental Science in Australia: Acknowledging First Nations People and Country," Emilie Ens and her colleagues in the Cross-Cultural Ecology and Environmental Management Lab at Macquarie University consider the current interaction of Indigenous and Euro-American ecological sciences in Australia. As mentioned at the outset of this introduction, Indigenous Australians have intricately managed their traditional estates since time immemorial. However, with the onset of European colonization, Indigenous knowledge systems were ignored, suppressed, or exploited to support settler agendas (i.e., to survive an environment that was harsh and unfamiliar to European colonists). Notwithstanding, as the mainstream discourse around race, social justice, and inclusion continues to evolve, so too does the rhetoric of science in Australia, which is shifting to one more inclusive of Indigenous science and knowledge. Ironically, as Euro-American ecologists now struggle to cope with climate change in their own countries, they are increasingly turning to Australian Aboriginal custodians as experts in climate adaptation. This chapter finishes by pointing to new cross-cultural approaches that seem to be working to open up space for more globally inclusive scientific methodologies and epistemologies.

In chapter 3, "African Sciences and Indigenous Knowledge Systems in the West African Ebola Crisis," Toluwani Oloke and Olusegun Soetan analyze the problems that Euro-American medical practitioners and agencies encountered in trying to get West Africans to adopt their treatment regimes. West Africans preferred to seek treatment from traditional healers, and Ebola fatality and recovery statistics were vastly underreported as a result. Based on their personal experience working in this crisis, and in the Yoruba culture, the authors conclude that the difficulties in coordinating Ebola efforts lay in the rhetorical failures of Euro-Americans to recognize scientific literacies among the Yoruba and other people, and to appreciate the distribution of African sciences throughout traditional religious, social, and mythical/sociolinguistic paradigms. The authors describe the African sciences and Indigenous knowledge systems among the Yoruba and explain how these belief systems influence the relationships and responses of Africans to health interventions in global health pandemics and epidemics. The authors conclude that a people's long-held cultural beliefs and practices cannot be alienated from that people during a crisis; in fact, the crisis in Nigeria served indirectly to promote and legitimate traditional African medicines as solutions to antibiotic resistance and other limitations of Euro-American medicine.

By contrast, in chapter 4, "A Critical Contextualized Approach to Studying Clashing Risk Cultures: Mapping the Transcultural Environmental Risk Communication of PM2.5 in China," Huiling Ding and Jianfen Chen treat a case in which Euro-American scientific standards were ironically recruited by local activists to combat Chinese technocracy. Identified as the culprit of smog plaguing many Chinese cities in winter, Particulate Matter (PM) 2.5 has become a household term since 2011 due to the nationwide debate on China's air quality standard. This chapter investigates the grassroots, networked risk communication and citizen-science endeavors around PM2.5 in China and the rhetorical effects of such arguments on China's air policies through the lenses of actor network theory and transcultural risk communication. The analysis reveals that the policy consensus on PM2.5 was achieved via an *ad hoc* collaborative intervention initiated by one environmental non-governmental organization (ENGO) on social media. The chapter concludes with a discussion of strategies for adding cultural nuance to cross-media analysis in transnational technical communication contexts in order to capture dynamics that run counter to established Euro-American communication norms.

Metaphor: Comparison as an Engine of Reinvention

Literal volumes have been written about the role of metaphor in scientific practice.[48] From them we have learned that scientific models are in essence elaborated metaphors, that scientists rely on metaphor to move from domains they understand to unknown domains, and that the choice of metaphor in science communication can profoundly impact the public understanding of science. As it does in poetry, metaphor serves scientific inquiry by suggesting new comparisons between the known and the unknown, new ways to see, know, and talk about the world. This is perhaps the most powerful way in which global sciences are currently influencing the practice of Euro-American sciences—by suggesting alternative theories and practices, particularly in vulnerable communities and in the Global South, which is increasingly the battleground of climate change. In chapter 5, "Where Voyaging Ends: Social Cosmology on Rapa Nui," Francisco Nahoe suggests that Euro-American rhetorics of science can be helpfully challenged and expanded by considering the case of the famous *moai* of Rapa Nui (Easter Island). These monumental statues have long presented a challenge to Euro-American archaeology. Nahoe, a Rapa Nui citizen and descendant of a prominent archaeologist working on the island, proposes a novel interpretation of the *moai* as a "quasi-discourse," a configuration that simultaneously constructed physical and social cosmologies on the island. The building of the *moai* appears to have coincided with the end of long-range navigational culture of on Rapa Nui and is thus best read, according to Nahoe, as a continuation of Polynesian ethnoastronomy. He argues that *moai* production both conserved the ethnoastronomy of navigation and channeled it into another social project: maintaining Polynesian social identity and cosmology in the absence of being able to voyage to connect with other Polynesian people groups. Nahoe concludes that appreciating rhetoric as social cosmology expands the capacity of ROS to work with global sciences that transmit knowledge in ways that exceed traditional Euro-American discursive norms.

In chapter 6, "Celtic Geometric Art as a Visual Rhetoric of Science," Evelyn Dsouza investigates the inventive potential of mathematical concepts developed, and overlooked, in the shadow of Anglo-American mathematics. A recent breakthrough polymerization technique in Euro-American chemistry was inspired by Celtic knots, complete loops without a beginning or end that have adorned religious monuments and manuscripts since the time of the late Roman Empire. By delving into their history, Dsouza learns that Celtic knotwork reveals the deep mathematical sophistication of their creators, an Indigenous knowledge suppressed by the colonization of Ireland and its racist policies. By way of rhetorical sequencing, an analytical technique in historical research, this chapter recovers a Celtic rhetoric of mathematics and ecology as made manifest in the mathematical and artistic figuration of the knot. The chapter first attempts to understand Celtic knots on their own terms, as figuring the nexus between nature and culture, and then examines their uptake in current scientific and mathematical thought—imagining them as an alternative to traditional figurations in Anglo-American rhetorics of science, one that may be perfectly poised to advance complexity science and ecology. These chapters invert Eve Tuck and K. Wayne Yang's critique of decolonization as a metaphor in order to examine the actual social and material impacts of scientific metaphors in colonial contexts.[49] Taken together, they suggest that engaging Indigenous terms of comparison and reference in these contexts will necessarily change the relationship between science and society.

Metonymy: Telling Non-Reductive Stories

Metonymy is closely related to synecdoche in that it provides a handle to grasp when confronting a large, complex idea or problem. But where synecdoche provides a part of the whole to grasp, metonymy offers an association—a linked symbol or emblem to discuss in place of the subject at hand.[50] For example, "the Crown" serves as a handy metonym for the tangled genealogical and political history of the Windsors, and we say someone "took the badge" instead of narrating the complicated process by which they became a law enforcement officer. Importantly, stories themselves function metonymically, to explore philosophical concepts or teach social norms indirectly; these are our myths and morality tales. By linking events and concepts into associative configurations rather than reducing them to a few generative causes like "society" or "patriarchy," metonymic narration offers a powerful alternative to traditional Euro-American scientific narratives, one that makes space for a diversity of voices. For this reason, feminist scholars of composition and technical communication have embraced metonymy.[51] We have at least two strong examples of metonymic rhetorical invention in the volume.

Chapter 7, "This is a Viral Story about Viral Stories: Image and Graphical Power in COVID Communication in the Navajo Nation," by Sunnie R. Clahchischiligi, Julianne Newmark, and Joseph Bartolotta, examines multimodal and media-rich ways of communicating about COVID-19 in Tuba City, Arizona, that foreground Diné (Navajo) traditions for understanding and talking about health and illness. The Navajo Nation achieved astounding rates of vaccination, nearly 80 percent, largely through internal cultural messaging. In investigating this case, the authors embrace "storying" as an Indigenous rhetorical method that serves as a powerful and effective counterpoint to traditional technical communication. They consider a number of Indigenous media messages about COVID-19, first via a story by Clahchischiligi, a member of the Navajo Nation, and then afterward by traditional rhetorical analysis: the two accounts stand side-by-side and illuminate each other without insisting that inconsistencies or excesses be reconciled. This approach creates a richer account of technical communication in a vulnerable community than a traditional Euro-American analytical approach could create.

Finally, in chapter 8, "A Rhetoric of the Home Ground: Local Knowledge and Data-Gathering among the North Atlantic Glaciers," Ryan Eichberger imagines another way to hybridize Euro-American and Indigenous sciences without reducing one to the other. Eichberger uses the Icelandic glacier as a lens through which to view not only the development of a specific Euro-American science—glaciology—but also a contemporaneous Indigenous tradition for understanding and living with glaciers. Early settlers recorded vivid verbal-visual impressions of the glaciers in classic works of Icelandic literature, such as *Egil's Saga*, *Grettir's Saga*, and the *Book of Settlements*. Then, beginning in the Enlightenment era, the Danish natural philosophers who indexed Iceland for purposes of colonial governance made their own technical drawings. Glaciers were visually appropriated by 20th and 21st century glaciology, which rendered them as satellite maps, thermal gradients, and ice-core strata; local photographers also made their own records of retreating glaciers. The glaciers thus exist in different ways for different communities at different times, and these visualizations sometimes clash as Icelanders decide how to live with their melting heritage. Eichberger attempts to set all these different glaciers side by side in his narrative, which both complicates our understanding of the development of Euro-American science in Iceland and suggests that paradigms pitting Euro-American sciences against Indigenous knowledges will be insufficient to frame moments of cultural contact and transition in vulnerable locations.

As should be apparent from the above synopsis, this volume on global rhetorics of science comprises a diverse array of contexts, methods, voices, and styles. Most of the scholars in this volume do not even identify as rhetoricians. Notwithstanding, all share a commitment to better understanding the relationship between world-building and community-building, and to contributing to the decolonization of scientific practice around the world. As they engaged in this project of reconfiguring global rhetorics of science, they attempted to observe key principles set out by the comparative and decolonial rhetorical scholarship reviewed above. They wrote about communities in which they lived or belonged to whenever possible, and when this wasn't possible, they gave the community space to speak for itself, engaging rhetorics of listening from close readings of social media to interviews.[52]

It is my hope that readers will come to our volume with an open mind. Some of our chapters do not sound like traditional academic arguments; some present ideas that may be uncomfortable or even shocking to ROS scholars. But I would argue that we cannot hope to diversify the "science" in "rhetoric of science" if we are not willing to change the way we understand and talk about science in the first place. The contributors to this volume help us take a first step on that path, and for that gift, as their editor, I am more grateful than I can say. I want to close this introduction by extending our work as an invitation to like-minded scholars to put their shoulders to the wheel of a more equitable and collaborative role for Euro-American science in the management of global risks. There is certainly no time like the present.

Notes

1. Leah Tonino, "Two Ways Of Knowing: Robin Wall Kimmerer On Scientific And Native American Views Of The Natural World," *Sun Magazine*, April 2016, https://www.thesunmagazine.org/issues/484/two-ways-of-knowing.

2. We have chosen to use the descriptor *Euro-American* for the current dominant global scientific tradition, instead of Western, to add more specificity to the descriptor and also to defuse the Orientalism resulting from West/East and Western/non-Western.

3. Steve Morton, Mandy Martin, Kim Mahood, and John Carty, eds., *Desert Lake: Art, Science and Stories from Paruku* (Collingwood, Australia: CSIRO, 2013), 23.

4. See, especially on "cosmic incommensurability," Randy Allen Harris, *Rhetoric and Incommensurability* (Chicago: Parlor Press, 2005).

5. Helaine Selin, *Encyclopaedia Of The History Of Science, Technology, and Medicine in Non-Westen Cultures* (Berlin: Springer Science & Business Media, 2013) this is the key reference on global sciences; also see Paul Keyser, *The Oxford Handbook of Science and Medicine in the Classical World* (Oxford: Oxford University Press, 2018). On mathematics, see Ubiratan D'Ambrosio, *Mathematics across Cultures: The*

History of Non-Western Mathematics, vol. 2 (Berlin: Springer Science & Business Media, 2001). For a defense of the term *global science* as used in this volume (as opposed to *nonwestern science* and *ethnoscience*), see Sujit Sivasundaram, "Sciences and the global: on methods, questions, and theory," *Isis* 101, no. 1 (2010): 146–58. For the impact of colonial imperialism on the development of Euro-American science, see G. Dawson B. V. Lightman, M. Elshakry, and S. Sivasundaram, *Victorian Science and Literature: Science, Race and Imperialism* (London: Pickering & Chatto, 2012). The special issue of *Isis* on global sciences (vol. 101) is excellent. Although "ethnoscience" is a contested term, much important work on global sciences has gone on under that heading: see in particular Stephan Rist and Farid Dahdouh-Guebas, "Ethnosciences—A Step Towards the Integration of Scientific and Indigenous Forms of Knowledge in the Management of Natural Resources for the Future," *Environment, Development and Sustainability* 8, no. 4 (2006): 467–93; William C. Sturtevant, "Studies in Ethnoscience 1," *American Anthropologist* 66, no. 3 (1964): 99–131. A complete list of literature on individual global sciences beyond what is contained in the reference listed above is infeasible, but for major and/or classical traditions: On African sciences, G. Emeagwali and G. J. S. Dei, *African Indigenous Knowledge and the Disciplines* (Rotterdam: SensePublishers, 2014). For Islamic science, see Osman Bakar, *Tawhid and Science: Essays on the History and Philosophy of Islamic Science* (Penang, Malaysia: Secretariat for Islamic Philosophy and Science, 1991); George Saliba, *Islamic Science and the Making of the European Renaissance* (Cambridge, MA: MIT Press, 2007). For science and mathematics in pre-Columbian American societies, see James J. Aimers and Prudence M. Rice, "Astronomy, Ritual, and the Interpretation of Maya 'E-Group' Architectural Assemblages," *Ancient Mesoamerica* 17, no. 1 (2006): 79–96; Marcia Ascher and Robert Ascher, *Code of the Quipu: A Study in Media, Mathematics, and Culture* (Ann Arbor: University of Michigan Press, 1981); Francisco Guerra, "Aztec Science and Technology," *History of Science* 8, no. 1 (1969): 32–52; Robin Wall Kimmerer, *Braiding Sweetgrass: Indigenous Wisdom, Scientific Knowledge and the Teachings of Plants* (Minneapolis, MN: Milkweed Editions, 2013). The majority of solid work on Classical Chinese sciences has been published in Mandarin and Russian, but see the Science and Civilisation in China series from Cambridge University Press, and on Classical Chinese medicine, K. Chimin Wong and Lien-teh Wu, *History of Chinese Medicine. Being a Chronicle of Medical Happenings in China from Ancient Times to the Present Period* (Tientsin: Tientsin Press, 1932).

6. Sivan Kartha, "Discourses of the Global South," in *The Oxford Handbook of Climate Change and Society*, ed. John S. Dryzek, Richard B. Norgaard, and David Schlosberg (Oxford: Oxford University Press, 2011), 504–19; J. Martínez-Alier, *The Environmentalism of the Poor: A Study of Ecological Conflicts and Valuation* (Cheltenham, UK: Edward Elgar Publishing, 2003). See both of these sources for general problems with climate justice in the Global South. For specific issues regarding scientific and technical communication, see the special volume of

Connexions on this topic, particularly Gerald Savage and Godwin Agboka, "Guest Editors' Introduction to Special Issue," *Professional Communication, Social Justice, and the Global South* (2016): 3.

7. Lynda Olman and Danielle DeVasto, "Hybrid Collectivity: Hacking Environmental Risk Visualization for the Anthropocene," *Communication Design Quarterly* 8, no. 4 (2020): 18–28.

8. In this volume we will use the term *communication* not to refer to the narrow field of communication studies but to a shared human practice of building the world and community simultaneously via *symbolic action* as defined by Kenneth Burke, *Language as Symbolic Action: Essays on Life, Literature, and Method* (Berkeley: University of California Press, 1966). I include material and affectual circulation under symbolic action because the action of signs, things, and energy is manifestly joint and not worth untangling for our present purpose, which is to introduce the new field of global rhetorics of science. Further work will doubtless wish to untangle those modalities of communication in particular situations.

9. George A. Kennedy, *Aristotle on Rhetoric: A Theory of Civic Discourse: Translated with Introduction, Notes and Appendices* (Oxford: Oxford University Press, 2007), 37.

10. Yasmeen Mahnaz Faruqi, "Contributions of Islamic Scholars to the Scientific Enterprise," *International Education Journal* 7, no. 4 (2006): 392-3.

11. Saliba, *Islamic Science*, 66.

12. Faruqi, "Contributions of Islamic Scholars," 393–94; Saliba, *Islamic Science*, 193–4; El-Sayed El-Bushra and M. M. Muhammadain, "Perspectives on the Contribution of Arabs and Muslims to Geography," *GeoJournal* 26, no. 2 (1992): 157–66.

13. Jim Al-Khalili, *The House of Wisdom: How Arabic Science Saved Ancient Knowledge and Gave Us the Renaissance* (London: Penguin, 2011).

14. See for instance Marwa S. Elshakry, "Knowledge in Motion: The Cultural Politics of Modern Science Translations in Arabic," *Isis* 99, no. 4 (2008): 701–30; chapters 14 to 19 in Michael J. L. Young, John Derek Latham, and Robert Bertram Serjeant, *Religion, Learning and Science in the 'Abbasid Period* (Cambridge: Cambridge University Press, 2006); and chapter 2 in Saliba, *Islamic Science*.

15. Patrick Kingsley, "Turkey Drops Evolution from Curriculum, Angering Secularists," *New York Times*, April 23, 2017, https://www.nytimes.com/2017/06/23/world/europe/turkey-evolution-high-school-curriculum.html.

16. Seng Loo, "Islam, Science and Science Education: Conflict or Concord?," *Studies in Science Education* 36 (2001): 45–77, https://doi.org/10.1080/03057260108560167; Suleman Dangor, "Islamization of Disciplines: Towards an Indigenous Educational System," *Educational Philosophy and Theory* 37, no. 4 (2005): 519–31, https://doi.org/10.1111/j.1469-5812.2005.00138.x.

17. Eric Winkel, "Tawhw and Science: Essays on the History and Philosophy of Islamic Science, by Osman Bakar (Review)," *Muslim World* LXXXIII, no. 3–4 (1993): 329–35.

18. Bakar, *Tawhid and Science*, 1-2.

19. Kenneth Burke, *Language As Symbolic Action*; Chaim Perelman and Lucie Olbrechts-Tyteca, *The New Rhetoric* (Notre Dame, IN: Notre Dame University Press, 1971).

20. Ulrich Beck, Scott Lash, and Brian Wynne, *Risk Society: Towards a New Modernity*, (London: Sage, 1992).

21. G. Thomas Goodnight, "The Personal, Technical, and Public Spheres of Argument: A Speculative Inquiry into the Art of Public Deliberation," *The Journal of the American Forensic Association* 18, no. 4 (1982): 215.

22. This work on networks and topologies in rhetoric has been inspired by the post-critical and spatial turns, particularly by the work of Gilles Deleuze, Félix Guattari, and Bruno Latour. For examples, see these edited volumes: Paul Lynch and Nathaniel Rivers, *Thinking with Bruno Latour in Rhetoric and Composition* (Carbondale, IL: SIU Press, 2015); Lynda Walsh and Casey Boyle, *Topologies as Techniques for a Post-Critical Rhetoric* (New York: Springer, 2017).

23. This work on materialist rhetorics has two major strands: Marxist and Heideggerian. A good example of a Marxist approach can be found in Kelly Happe's chapter in this volume (chapter 5). A good example of a Heideggerian approach to materialist rhetorics is Thomas Rickert, *Ambient Rhetoric: The Attunements of Rhetorical Being* (Pittsburgh, PA: University of Pittsburgh Press, 2013).

24. LuMing Mao, "Doing Comparative Rhetoric Responsibly," *Rhetoric Society Quarterly* 41, no. 1 (2011): 64-69. See also Scott R. Stroud, "Pragmatism and the Methodology of Comparative Rhetoric," *Rhetoric Society Quarterly* 39, no. 4 (2009): 353-79; Bo Wang, "Comparative Rhetoric, Postcolonial Studies, and Transnational Feminisms: A Geopolitical Approach," *Rhetoric Society Quarterly* 43, no. 3 (2013): 226-42.

25. On contact zones see Mary Louise Pratt, "Arts of the Contact Zone," *Profession* (1991), http://www.jstor.org/stable/25595469. For works setting out the terms of decolonial studies of rhetoric in general, see Lisa Flores, "Advancing a Decolonial Rhetoric," *Advances in the History of Rhetoric* 21, no. 3 (2018): 320-22; Romeo García and Damián Baca, "Rhetorics Elsewhere and Otherwise: Contested Modernities, Decolonial Visions" (Champaign, IL: NCTE, 2019); Amardo Rodriguez, "A New Rhetoric for a Decolonial World," *Postcolonial Studies* 20, no. 2 (2017): 176-86. For a critical/decolonial collection of American rhetorics, see Damián Baca and Victor Villanueva, eds., *Rhetorics of the Americas: 3114 BCE to 2012 CE* (New York: Palgrave Macmillan, 2010).

26. Malea Powell, "Rhetorics of Survivance: How American Indians Use Writing," *College Composition and Communication* 53, no. 3 (2002): 398.

27. Ellen Cushman, "Translingual and Decolonial Approaches to Meaning Making," College English 78, no. 3 (2016): 236.

28. Lisa King, Rose Gubele, and Joyce Rain Anderson, *Survivance, Sovereignty, and Story: Teaching American Indian Rhetorics* (Boulder: University Press of Colorado, 2015), 4.

29. Randy Allen Harris, *Landmark Essays on Rhetoric of Science Case Studies* (Mahwah, NJ: Lawrence Erlbaum Associates, 1997). See this source for a good sample of first-wave ROS studies. For the critical turn in ROS, see Alan G. Gross and William M. Keith, *Rhetorical hermeneutics: Invention and interpretation in the age of science* (Albany, NY: SUNY Press, 1997).

30. On intersectional issues around race, class, and gender in ROS see Celeste M. Condit, "How the Public Understands Genetics: Non-Deterministic and Non-Discriminatory Interpretations of the "Blueprint" Metaphor," *Public Understanding of Science* 8, no. 3 (1999): 169–80, https://doi.org/10.1088/0963-6625/8/3/302; Kelly E Happe, *The Material Gene: Gender, Race, and Heredity After the Human Genome Project* (New York: NYU Press, 2013); Lisa Keränen, *Scientific Characters: Rhetoric, Politics, and Trust in Breast Cancer Research* (University of Alabama Press, 2010); James Wynn, *Citizen Science in the Digital Age: Rhetoric, Science, and Public Engagement* (Tuscaloosa: University of Alabama Press, 2017). On disability in ROS see Jordynn Jack, *Autism and Gender: From Refrigerator Mothers to Computer Geeks* (Champaign: University of Illinois Press, 2014); Jenell Johnson, *American Lobotomy: A Rhetorical History* (Ann Arbor: University of Michigan Press, 2014). On nonhuman agency and posthumanism see S. Scott Graham, *The Politics of Pain Medicine: A Rhetorical-Ontological Inquiry* (Chicago: University of Chicago Press, 2015); Kristen R. Moore and Daniel P. Richards, *Posthuman Praxis In Technical Communication* (New York: Routledge, 2018).

31. Selin, *Encyclopaedia*.

32. Miriam F. Williams, "Reimagining NASA: A Cultural and Visual Analysis of the U.S. Space Program," *Journal of Business and Technical Communication* 26, no. 3 (2012): 368–89.

33. Julianne Newmark, "The Formal Conventions of Colonial Medicine: Bureau of Indian Affairs' Agency Physicians' Reports, 1880–1910," *College Composition and Communication* 71, no. 4 (2020): 620–42; Kenneth C. Walker, *Climate Politics on the Border: Environmental Justice Rhetorics* (Tuscaloosa: University of Alabama Press, 2022).

34. Godwin Y. Agboka, "Decolonial Methodologies: Social Justice Perspectives in Intercultural Technical Communication Research," *Journal of Technical Writing and Communication* 44, no. 3 (2014): 297–327.

35. Angela M. Haas, "Race, Rhetoric, and Technology: A Case Study of Decolonial Technical Communication Theory, Methodology, and Pedagogy," *Journal of Business and Technical Communication* 26, no. 3 (2012): 277–310. Also, see the special issue of the *Journal of Business and Technical Communication* (24, no. 4) and Cana U. Itchuaqiyaq's *Decolonial Methods in TPC: FULL Undergrad Course* (2020), https://www.itchuaqiyaq.com/post/decolonial-methods-in-tpc-full-undergrad-course.

36. Henry Etzkowitz and Loet Leydesdorff, "The Dynamics of Innovation: From National Systems and 'Mode 2' to a Triple Helix Of University–Industry–Government Relations," *Research Policy* 29, no. 2 (2000): 109–23.

37. David Cutler and Francesca Dominici, "A Breath of Bad Air: Cost of The Trump Environmental Agenda May Lead to 80,000 Extra Deaths Per Decade," *JAMA* 319, no. 22 (2018): 2261–62; Kari De Pryck and François Gemenne, "The Denier-in-Chief: Climate Change, Science And the Election of Donald J. Trump," *Law and Critique* 28, no. 2 (2017): 119–26.

38. On the effects of neoliberal policies on the health of Black women in the US, see Happe, *Material Gene*; on the displacement of Malaysian fishing communities as the result of transnational "climate services," see Helga Leitner and Eric Sheppard, "From Kampungs To Condos? Contested Accumulations Through Displacement in Jakarta," *Environment and Planning A: Economy and Space* 50, no. 2 (2018): 437–56; on the disruption of food sovereignty in Nicaragua by transnational NGOs, see Birgit Müller, "The Temptation of Nitrogen: FAO Guidance for Food Sovereignty in Nicaragua" (paper presented at Food Sovereignty: A Critical Dialogue, International Conference, Yale University, CT, September, 2013), 14–15.

39. Hayden White, *Metahistory: The Historical Imagination in Nineteenth-Century Europe* (Baltimore, MD: Johns Hopkins University Press, 2014), 31–35.

40. Michael J. Horswell, *Decolonizing the Sodomite: Queer Tropes of Sexuality in Colonial Andean Culture* (Austin: University of Texas Press, 2005), 12–13; Joel Wainwright, *Decolonizing Development: Colonial Power and the Maya* (New York: John Wiley & Sons, 2011), 78–85.

41. Edward P. J. Corbett and Robert J. Connors, *Style and Statement* (New York: Oxford University Press, 1999), 61.

42. Kimmerer, *Braiding Sweetgrass*, 40–45. See also Shawn Wilson, *Research is Ceremony: Indigenous research Methods* (Halifax, NS: Fernwood Publishing, 2020), 50.

43. Pamela Pietrucci and Leah Ceccarelli, "Scientist Citizens: Rhetoric and Responsibility in L'Aquila," *Rhetoric and Public Affairs* 22, no. 1 (2019): 95–128.

44. Philip Sherwell, "$40bn to Save Jakarta: The Story of the Great Garuda," *The Guardian*, November 22, 2016, https://www.theguardian.com/cities/2016/nov/22/jakarta-great-garuda-seawall-sinking.

45. Graham, *Politics of Pain Medicine*, 144.

46. Corbett and Connors, *Style and Statement*, 69.

47. Jonathan Swift, "A Modest Proposal for Preventing the Children of Poor People in Ireland from Being a Burden on Their Parents or Country and for Making Them Beneficial to the Publick," Project Gutenberg, https://www.gutenberg.org/files/1080/1080-h/1080-h.htm.

48. See for example Ken Baake, *Metaphor and Knowledge: The Challenges of Writing Science* (Albany, NY: SUNY Press, 2003); Max Black, *Models and Metaphors* (Ithaca, NY: Cornell University Press, 2019); Theodore L. Brown, *Making Truth: Metaphor in Science* (Chicago: University of Illinois Press, 2003); Thomas S. Kuhn, "Metaphor in Science," in *Metaphor and Thought*, ed. Andrew Ortony (London: Cambridge University Press, 1979), 533–42.

49. Eve Tuck and K. Wayne Yang, "Decolonization Is Not a Metaphor." *Tabula Rasa* 38 (2021): 61–111.

50. Corbett and Connors, *Style and Statement*, 62.

51. Joy Ritchie and Kathleen Boardman, "Feminism in Composition: Inclusion, Metonymy, and Disruption," *College Composition and Communication* 50, no. 4 (1999): 585–606.

52. Criteria drawn from Agboka, "Decolonial Methodologies"; Mao, "Doing Comparative Rhetoric Responsibly"; LuMing Mao et al., "Manifesting a Future for Comparative Rhetoric," *Rhetoric Review* 34, no. 3 (2015): 239–74; Darrel Wanzer-Serrano, "Decolonial Rhetoric and a Future Yet-to-Become: A Loving Response," *Advances in the History of Rhetoric* 21, no. 3 (2018): 326–30.

Chapter One

How Euro-American Science Became Dominant
Transnational Circulations of Knowledge and Capital

Kelly E. Happe and Lynda C. Olman

With COVID-19 vaccines being manufactured and distributed by large pharmaceutical companies, seemingly all of them named after their corporate sponsors,[1] there is perhaps no better time to consider the relationship between science and capitalism. What is particularly striking about current events is the contrast between the resources put into developing the vaccine and a global public health context in which most people will have to wait a long time to get vaccinated (if they get vaccinated at all), much less be able to live in conditions less conducive to the virus's spread.[2] What COVID-19 demonstrates in disheartening detail is that global capitalism, and Euro-American dominance in science and technology, mean that a lack of even the most basic public health infrastructure—compounded by extreme poverty—ensures both the creation of health crises and the justification for narrow, corporatized approaches to solving them.

Despite the fact that it has long been shown that profit is neither a necessary nor sufficient condition for innovation in science and medicine (Cuba, for example, has a history of developing innovative, successful vaccines against the flu),[3] it is nevertheless the case that we live in a world in which scientific research and technological innovation are inextricably linked with the profit logic of capital. Kaushik Sunder Rajan calls this nexus the coproduction of science and capital, a global phenomenon in which the needs, interests, and desires of both converge on a transnational scale.[4] On the one hand, the history of this coproduction is one of raw power, as expropriation cleared the way for both the spread of capitalism and the knowledge industries on which it increasingly depends.[5] On the other, it is a history of symbolic reduction, also enabled by expropriation, insofar as Euro-American science, in which the value of innovation (and knowledge itself) is really exchange value, has come to stand in for all science. For example, scholars such as Vandana Shiva, Richard Lewontin, and Richard Levins have documented both material dispossession and symbolic reduction in the case of agriculture, in which Euro-American capitalist methods for industrial monoculture not only wreak havoc on food production but also violently displace local practices around farming—all of which has been justified through ideological framings that privilege white, Euro-American, and masculinist science and its technocratic imaginaries.[6]

More recently, we see this coproduction in the case of egg vending for stem cell research and other kinds of "biomedical labor" (e.g., participation in clinical trials) in which bodies (their labor and their materiality) are caught up in global circuits of capital accumulation. Not only does this labor mark an instance of neocolonial expropriation and exploitation—poverty, debt, and unemployment produce laborers who can be exploited in altogether new ways—it is also an example of the differential value of labor in biocapitalism in which the "intellectual" labor of the scientist is distinct from the "corporeal" labor of the egg donor or research subject. This bifurcation then allows for the devaluing of the latter, if in fact the latter is recognized as "labor" at all. All of this is part and parcel of what Sunder Rajan calls *biocapitalism*: the emergence of economies in which the body as such is reduced to its economic value, whether that be through logics of commodification or speculative finance.[7]

We argue in this chapter that this symbolic reduction of all knowledge, expertise, labor, and embodiment to economic value is an essentially *rhetorical* process, having been negotiated at specific historical times and places by specific agents. To be more exact, it is a *synecdochal* process. In other words, the understanding of science as always already a Euro-American phenomenon did not spread globally because of its inherent virtues but because of the material and symbolic violence that was a condition of its circulation. Thus, an important question for scholars interested in the relationship between science and capitalism is not only how they are currently imbricated in various ways, but also how what we call modern science—an Anglo-European designation—became certified as the only legitimate mode of global knowledge-making. The spread of capitalism certainly explains this dynamic, insofar as science and capitalism share epistemic structures (e.g., logics of "innovation," "discovery," and "speculation," to name just a few); then, the spread of one will presume the spread of the other. However, attending more closely to the ways in which Euro-American science was argued into place as the dominant global ideo-epistemology will allow rhetoric of science scholars to consider not just what was lost along the way, but crucially, what might be reclaimed, albeit under material and symbolic conditions vastly different than what came before.

Euro-American Science as Global Force

If we think of science as having a history, and one that is inextricably tied to the material conditions in which it is practiced, then what we mean by Euro-American science itself has changed as well. Of course, we already know science has a "history," an idea put into circulation by Thomas Kuhn's well-known book *The Structure of Scientific Revolutions*, in which he laid out the theory of paradigm shifts.[8] Taking physics as an exemplar, Kuhn argued that the shift from Newtonian to quantum physics was so profound that one could say that physicists were not practicing within the same world as one another. The only explanations that could suffice to explain the move from one paradigm to another were aesthetic—some scientists were simply drawn to Einstein's theories, in some cases through an abrupt gestalt switch—or material, as when older scientists simply lost resources and power.

Whether one agrees with Kuhn's central argument about the incommensurability of paradigms or not, what he showed was that change occurs not because of the evolution of rational thought but from the rearrangement of material resources. These changes, however, were described by Kuhn as largely internal to science. The approach we take here is congruent with that taken by other scholars who see science as part of a social totality, which is to say, as part of the dominant mode of production and the kinds of relations it enables and makes intelligible and desirable. Importantly, what these scholars have also emphasized is that, when we talk of the relationship between science and capitalism, we must acknowledge its always already global dimensions.

The mutual development of science and capitalism is multifaceted but can be generally categorized under the conceptual heading *coproduction*, which understands that science and capitalism have distinct practices that are contingent, cannot necessarily be anticipated in advance, and are analytically distinct—even if, as Sunder Rajan maintains, they are ultimately governed by market logics.[9] Put another way, although science makes value for capital in some way, we nevertheless need to think about *how* scientific and capitalistic interests converge and, keeping with the insights of historical materialism, how that convergence changes over time. We posit here, drawing on Sunder Rajan and other scholars, that coproduction is the result of the following factors: epistemological commensurability; the naturalization of class interests; the differential value attached to labor; and the transformation of life itself into exchange value. All of these factors are enabled by and in turn enable the expropriation of land and knowledge and the capitalization of the body.

The earliest stages of merchant capitalism, for instance, laid the groundwork for such a convergence in an increasingly globalizing and imperializing context.[10] Writing about the coproduction of "commercial materialism and philosophical materialism" (the latter a precursor to modern science), Harold Cook claims,

> As for the European Scientific Revolution: where merchants gained power so did the kinds of commensurable material knowledge they most valued. In a growing number of places, merchants governed the cities that depended on their activities; they also managed to engineer city-states and republics. Even where they could not gain sovereignty, however, they became indispensable to the rulers of principalities and kingdoms by lending them the ready money required to exercise power . . . By the later seventeenth century, a class of rentiers can be identified whose wealth derived chiefly from investment instruments . . . who added modeling and the channeling of energy to their interest in descriptions and calculations of material and space. As merchants and capitalists began to make robust claims on the future of nations, the kind of knowledge processes they had long valued grew in prominence, opening doors to what would come to be called modern science.[11]

Thus, the norms of science reflect what would come to be class interests, which is to say, science and capital share material interests and abstract logics because what we know to be Euro-American science is coincident with the rise of the bourgeoisie in the early modern period. For Levins and Lewontin, what we call the "Cartesian" method is an ideology[12] whose "ontological commitments" are characterized by a particular understanding of the relationship between parts and wholes, cause and effect[13]—in other words, a worldview that is reductionist and, in the Marxist lexicon, "alienated":

> We characterize the world described by these principles as the *alienated* world, the world in which parts are separated from wholes and reified as things in themselves, causes separated from effects, subjects separated from objects. It is a physical world that mirrors the structure of the alienated social world in which it was conceived. Beginning with the first glimmerings of merchant entrepreneurship in thirteenth-century Europe, and culminating in the bourgeois revolutions of the seventeenth and eighteenth centuries, social relations have emphasized the primacy of the alienated individual as a social actor. By successive acts of enclosure, land was alienated from the peasant cultivators, who formerly were tied to it and it to them. Individuals became social atoms, colliding in the market, each with his or her special interests and properties intrinsic to their roles.[14]

Levins and Lewontin thus show how otherwise abstract notions of "scientific method" are actually shot through with material class interests, which is also an explanation for how and why, in revolutionary contexts, science can be a counterrevolutionary force.[15] They are especially attentive to the role of expropriation in this process of global dominance of bourgeois science, not only in the early stages of capitalism and its spread, but later, when industrialized for-profit agriculture was forcefully exported transnationally. For Levins and Lewontin, agriculture exemplifies the ways in which Euro-American capitalism and science have reigned triumphant and with devastating social and ecological effects. Indeed, the story of global agriculture demonstrates how developments in Euro-American science—namely biotechnology and the industrial farming methods it has enabled—are circulated, and that such circulation is enabled, constrained, and/or coerced by a global infrastructure of laws, regulations, and norms. Overcoming the "natural" limits to capital accumulation requires not only the dispossession of land but the usurpation of local farming practices.[16]

Edward Yoxen turns even deeper into the world of scientific practice to demonstrate the convergence of biology and capitalist interests, elaborating on what he calls "alignment of interest."[17] He shows not only how the supposed "autonomy" of the sciences was never a real bulwark against the entanglement of science and capitalism, but also how expropriation of creative labor power was part and parcel of developments within the sciences (to enable what is essentially epistemological commensurability) in the globalization of this new industry. In his essay "Life as a Productive Force," Yoxen begins with a history of how science became "directed" (managed) and scientists "reskilled" to serve both state and industry, for example when the US federal and industry funding of science, especially after World War II, guided the kinds of research projects scientists pursued.[18] For Yoxen, to understand the "assimilation or annexation of biology by capital" we need to look not only at how the interests of scientists and capital came together, but also how the myth of the autonomy of the sciences served to disavow said alignments. Moreover, since a range of political perspectives were reconciled to the shared aim of "capitalising life,"[19] Yoxen examines scientific "concepts, methodological assumptions, and the control of meanings which become attached to scientific ideas, to specify how scientists' activities and plans evade, dovetail with or catalyse this historical process of incorporation."[20]

Echoing Lewontin's argument that the transformation of living matter was essential to the increased corporate control of science here and abroad, Yoxen documents that, in addition to the use of recombinant DNA to breed new plants, there was a push to "bank" seeds and genes so as to create markets in which said commodities could be sold to the developing world.[21] He concludes,

> In this sense "biotechnology" is not simply a way of using living things that can be traced back to the Neolithic origins of fermentation and agriculture. As technology controlled by capital, it is a specific mode of the appropriation of living nature—literally capitalising life. It constitutes a new set of social relations between fundamental research scientists, industrial corporations, and state agencies. Its *telos* is to promote technology while regulating the deskilling and displacement of workers. It also serves to inhibit collective resistance on the part of "consumers" of health services, expropriated peasants and custodians of plant gene resources.[22]

Kean Birch and David Tyfield further theorize the role of labor in the capitalization of life by explaining how expropriation manifests in a contemporary context in which financialization and speculation are dominant drivers of a global biotech industry. Not only is labor a key (if not the key) source of value-making in the biosciences, but how it becomes part of exchange and circulation in the global bioeconomy also demonstrates the need for (a) a transnational focus and (b) attention to how knowledge, as an asset, makes value in extractive and deeply exploitative ways.[23] As an illustration, Birch and Tyfield point to gene patents as part of a "rentier" regime in which profit is generated from said (scarce) asset.[24] This regime is enforced by the coercive measures of entities like the World Trade Organization:

> Thus rent-seeking strategies have only assumed predominance because financialization had led to a wholesale transformation of political-economic priorities away from productive capital (i.e., commodity production) toward finance capital (i.e., asset ownership) . . . what this has meant is the opening up or colonizing by capital of the social spaces of knowledge production and, thence, natural or biological productivity . . . This colonization includes the enactment of a global IPR [intellectual property right] regime; the growing commercialization of academia; the appropriation of indigenous knowledges.[25]

Biotech thus provides a powerful example of the processes by which capitalism and science work together historically to reduce life and labor to capital, and global knowledge systems to Euro-American technoscientific expertise.

Transnational Science and Symbolic Reduction

What the previous section has shown is that the process by which Euro-American science has become dominant has been one of the global coarticulation of knowledge and capitalism.[26] Because the histories of science and capital have been so tightly intertwined, the spread, and eventual dominance, of Euro-American science has been coincident with the (often violent) expropriation of land, labor, and life. What we explore next is the role of symbolic reduction that is part and parcel of material expropriation, which is to say, the synecdochal process via which Euro-American science has come to stand in, rhetorically, for all science.[27] To investigate the rhetorical aspects of this process is not meant to diminish the role of material violence, only to elucidate the many dimensions of expropriation at work—dimensions that are otherwise undertheorized with the coproduction concept.

We argue that symbolic reduction is the result of the rhetorical production of value: the coproduction of science and capital by way of epistemological reduction; the reduction of the value of labor to that considered "cognitive" or "innovative"; and finally, the reduction of life itself to exchange value. Regarding the first—epistemology—Sunder Rajan draws from Marx, who, he says, was concerned with the role of political economy in making intelligible and, more importantly, *naturalizing* the working of capital. Political economy also cleverly succeeded in establishing itself as the authority on such matters. For Sunder Rajan, a particularly Marxist approach to coproduction is appropriate insofar as science and culture share epistemological assumptions and, like political economy before, science makes intelligible and naturalizes market logics. He writes,

> What Foucault does explicitly is what I have argued Marx does implicitly, which is to consider political economy as consequential not (just) because it is a political and economic *system* of exchange but because it is a foundational *epistemology* that allows us the very possibility of thinking about such as a system *as* a system of valuation. The biopolitical, then, does not just refer to the ways in which politics impact everyday life, or in which debates over life (such as, to take an evident example, over new reproductive technologies) impact politics, but rather points to the ways in which our very ability to comprehend "life" and "economy" in their modernist guises is shaped by particular epistemologies that are simultaneously enabled by, and in turn enable, particular forms of institutional structures.[28]

Sunder Rajan argues not only that the epistemological is what enables the values of science and capital to align, but that there is a decidedly symbolic aspect to this process. For instance, he writes about the "shared grammar" of speculation that is part and parcel of knowledge making and investment alike.[29] "As a driving force of capital," he says, innovation "draws to a large degree on technoscientific potential for value generation;

technoscientific potential for value generation is fundamentally enabled by a political-economic regime that provides incentives to innovate."[30] Thinking about this in a transnational context, Sunder Rajan shows how such "incentives to innovate" are in turn made possible because of evolving narratives of nation and belonging. In his comparative account of India and the United States, Sunder Rajan shows how ideologies of nation help biotech companies secure a foothold in materially diverse, highly localized contexts.[31] Similarly, Levins and Lewontin have written about the embrace of "developmentalism" in which key actors in the Global South embrace the logic and telos of reductionist technoscience.[32] Innovation as trope and affective economies (for example, those structured by ideologies of nation) thus show the power of rhetoric in effecting the symbolic reduction needed for Euro-American science to become dominant globally while making it appear as though this outcome is simply the result of progressive movement, from the putative "primitive" to the "civilized." Along the way, global sciences are evacuated of meaning so that Euro-American understandings of those terms can take hold, and, with those understandings, the class interests of Euro-American expertise and capital accumulation.

Synecdochal reduction is also the rhetorical mechanism by which differential valuations of labor inhere, another way in which Euro-American science becomes dominant by drawing a distinction between the intellectual labor of the Global North and the raw material of the Global South. For example, the staging of clinical trials around the world by multinational pharmaceutical companies, the sale and circulation of bodily tissues, and the procuring of reproduction services like surrogacy are all examples of unacknowledged, unrepresented biomedical and clinical labor that, exactly because it remains unrecognized *as* labor, produces value for capitalism. Kalindi Vora writes, for instance, about the classification of "non-innovative" knowledge work within tech industries located in India as the aftermath of the long-standing demarcation of imperial Euro-American industry and the colonized natural resources upon which it has depended—a paradoxical classification, to be sure, and only possible if Euro-American knowledge systems are permitted to stand in synecdochically for all knowledge.[33]

Euro-American science, an atomistic, instrumental understanding of the world, grounded in a logic of innovation cum speculation, thus takes hold globally with the spread of capitalism. Relatedly, what is considered valuable is attached to the exercise of cognitive or intellectual labor, a definitional demarcation that renders clinical trial participants,

IT workers, and their cohort mere raw material for its exercise. The distinction between the science of the Global North and the raw material of the Global South is, furthermore, a logic that informs the capture of life itself by capital—what Edward Yoxen calls "life as a productive force" and Sunder Rajan calls biocapital.

Yoxen notes the example of patenting: when genes could be classified as "unnatural" they could then be "ownable artefacts, whilst also being part of living nature."[34] Genes, disarticulated from "life," can paradoxically produce value for capitalism while appearing to merely represent biological matter. Stefan Helmreich describes this disarticulation as the performative production of value when biology is seen as endlessly generative. His analysis of a variety of texts, including biotech company reports, conference talks, mission statements of research centers, and the like, shows that the affective sentiments of biotech actors bring into being the very value they presume to already inhere in biological matter itself. Extending Marx's formula for capital (M-C-M', meaning that money turns into commodities that turn into profit, wherein the purpose of circulation is accumulation), Helmreich suggests B-C-B', where B stands for biomaterial and B' the surplus value generated by the biotech product. From this equation, he further suggests that "the sentiment of many biotech boosters has them taking B' already to be latent in B—to believe, that is, that biological process itself already constitutes a form of surplus value production."[35] The economic metaphors he finds in this discourse do not, as Helmreich rightly notes, reflect the reality of biological systems as much as they constitute biology as value-making in a way that can be seen as analogous to industrial capitalism. He writes, "The appearance of the *bio* in biocapital *as* reproductive is constituted by the capitalist enterprise that turns it into something that generates exchange-value in the first place."[36]

Implicit in Helmreich's account is the claim that, with no stable referent, biology can be developed for the production of exchange value, with exchange value standing in for *all* value. Developments in biology are thus a parallel case to what Donald Lowe theorized in his book *The Body in Late Capitalist USA*: that late capitalism augurs the "hegemony of exchangist values" in which sociocultural values—for example, those associated with bodily needs—are primarily understood as signifiers for exchange value.[37] In the case of biocapital, we see the destabilization of the biological itself, with risk displacing health and producing what Sunder Rajan calls "patients-in-waiting/consumers-in-waiting" in the Global North, a subject position that embodies speculation as a rationale. As we have established, this is a relational subjectivity, dependent upon the extraction of resources from elsewhere. Placing this dynamic in a context of the history of colonialism and resource expropriation, Vora calls biocapitalism the global transmission of "vital energy": the "substance of activity that produces life (though often deemed reproductive)—from areas of life depletion to areas of life enrichment."[38] What Vora describes is possible in part because "life," broken into so many vital fragments, becomes surplus to the body and so a source of surplus value—a transformation made possible because the biological models crafted to manufacture such surplus are themselves modes of symbolic action in the world.[39]

To further elaborate upon the extractive, expropriative, and, as we shall see, exploitative and symbolically reductive aspects of the science/capital nexus, we turn now in more detail to the case of stem cell research.

Material and Symbolic Dispossession at the Life Science/Capital Nexus: The Case of Stem Cells

Stem cell research and industry exemplify the ways in which biomedicine and capitalism alike are transformed as the body becomes the source of value: the value of health and longevity and the value of capital accumulation and profit. Animated by the promise of "regenerative" medicine, stem cell researchers hope to develop therapies (e.g., correcting for brain damage in stroke patients)[40] that can potentially be personalized (as when a patient's own stem cells are used to develop treatments). Whether or not stem cell research ever meets its more ambitious goals, it nevertheless helps illustrate what we have described as the material expropriation and symbolic reduction that characterize the globalization of Euro-American, capitalistic science.[41]

Biomedical research in this area depends on a variety of sources of stem cells, although embryonic stem cells have tended to be more useful to researchers since they are in a pluripotent state, meaning they are undifferentiated and thus have the potential to develop into a variety of tissues and organs. Donated embryos from fertility clinics are one source, although these may become damaged from preservation and storage techniques.[42] The number of oocytes available for stem cell research has remained insufficient, in part due to restrictions on selling ova for such purposes.[43] As a result, researchers have secured ova through transnational networks, targeting countries where regulations on such transactions are lax, if they exist at all.[44]

Given the transnational and transactional nature of stem cell research, Catherine Waldby and Melinda Cooper have elaborated on the ways in which it is inextricably bound up with localized histories of global capitalism. For example, those areas of the world in which egg vendors are most likely to reside are those that have historically been sites for the expropriation of land, labor, and other resources, as well as the crippling effects of debt. In these contexts, feminized laborers occupy new service labor roles (e.g., service work performed from the home) and also travel in search for service jobs elsewhere.[45] Leadership in these countries (e.g., in East and South Asia), moreover, actively seeks ways to participate in emerging bioeconomy markets either by incentivizing, outsourcing, or investing in biotechnology directly.[46]

Waldby and Cooper demonstrate that these developments require a new understanding of labor, one that can properly grasp the ways in which capital's historic need for generating new sources of value intersects with particular developments in the biomedical sciences. In the case of stem cell research, the very idea of surplus bodily material is possible because of new techniques for inducing what is called *super-ovulation* to generate manufactured excess for medical knowledge-making and, eventually, profit. This technoscience, moreover, transforms the concept of bodily potentiality, because the stem cells that are derived from the embryos these ova make possible can, in theory, be used to produce numerous cells, tissues, and organs. Put another way, the scientific idea of pluripotency wrests the stem cell from a singular developmental telos and places it squarely within the logics of speculative finance capital and its visions of endless accumulation.

These visions of profit are made possible by (even as they erase) the critical role that labor serves in making stem cell research possible. *Biomedical labor* is the term Waldby and Cooper use to describe the process by which value is produced when people are paid (or otherwise "compensated," as when donors are given money to cover the costs of travel, lodging, and such) to sell bodily material such as oocytes.[47] As they clarify, the labor involved in producing "surplus" bodily matter entails not only travel (sometimes extensive) but time and endurance, not to mention the aftereffects of such treatments, which can be debilitating or even deadly.[48] This labor, however, is made invisible in the neocolonial, hyper-exploitative contexts in which it is extracted. Waldby and Cooper write of this kind of embodied experience that "it does not consist primarily in the performance of codified tasks but rather in subjects giving clinics access to the productivity of their *in vivo* biology, the biological labor of living tissues and reproductive processes. It does, however, involve second-order tasks; compliance with often-complex medical regimes of dosing, testing, appointments and self-monitoring."[49] While this kind of labor is novel in many ways, Waldby and Cooper note that it bears traces of (and indeed, is unthinkable without) the global trade in enslaved persons and colonial expropriation, in which reproductive capacity was an asset and source of speculative value calculations. Indeed, ova vending is an example of the ongoing need of capital to incentivize and benefit from increased capacity of the body to produce surplus value (in this case the difference between what vendors are paid and the financial payoff of the research).[50] When we consider that the people most likely to benefit reside in the Global North (e.g., by benefitting from stem cell therapies), ova vending can be seen as another example of colonial, racial capitalism's extractivist logic.

What "value" means in science proper, as this example shows, is reduced to narrow definitions associated with the intellectual/cognitive capacity of the researchers, thus erasing the contributions made by ova vendors to the project of knowledge-making. Egg vendors are viewed as merely supplying raw material to an experimental regime that is made possible by the creative energies of scientists. Even as political economy theorists recast this cognitive labor as value-making in the capitalist sense (as Birch and Tyfield do), the value inherent in the biomedical labor of the donor/vendor is absented from the scholarly discourse. As Waldby and Cooper explain, "The organization of intellectual property in the life sciences recognizes the cognitive labor of the scientist and the clinician, but not the constitutive nature of the biological material or the collaboration of the donor. It is evident then that the recognition of labor here is structured by a mind/body split, wherein the embodied productivity of the tissue donor does not figure."[51] That this contribution is gendered—feminized, to be specific—makes it all the more difficult to acknowledge it as labor, especially since it is more likely to be seen as an altruistic "gift" than an embodied mode of labor exercised in a context of neocolonial expropriation[52]—a mystification enabled by a discourse of bioethics that has played a crucial part in making these emerging bioeconomies work.[53]

For there to be feminized "biomedical labor" at all, biology itself must be envisioned as producing profit. As we mentioned above, fertility research enabled the very idea of surplus biology through a technical innovation to increase production of eggs and their removal from ova-bearing persons. Further, by disarticulating pluripotency from logics of human reproduction and development, stem cells become conceptualized as a radical potentiality—a source of endless profit. As Waldby and Cooper observe,

> Formerly "reproductive tissue" enters into another epistemological space where the potentiality of the germ cell is defined in radically different ways. One of the prime innovations of stem cell science is to have reworked formerly orthodox understandings of cell potentiality. This is true of both somatic (non-reproductive) and germinal (reproductive) cells such as the egg or sperm. In each case, a notion of potentiality that formerly limited their future possibilities of division and differentiation to the evolving organism now detects a radically different, even incommensurable spectrum of possibilities in the same tissue specimen.[54]

They conclude, "Bodily potentiality is itself being reconfigured at the interface of new labor relations and the biological sciences."[55]

Stem cell research thus represents an exemplary case of Helmreich's argument that, in biocapitalism, biology is imagined as endlessly generative. What we wish to stress with respect to this case is the synecdochal process by which the stem cell becomes the site of innovation and speculation, two tropes that operate in qualitatively different ways in the sphere of science and the sphere of capital accumulation, but which nevertheless enable the stem cell to be reduced to exchange value. The technological developments whereby scientists have been able to isolate stem cells and, more radical still, induce pluripotency in already differentiated cells demonstrates the role of biological models themselves in capitalizing life. And as we discussed above, the very idea of capitalizing life is racialized and gendered, dependent on transnational circuits of resource extraction in which biology (in this case, the oocyte) is capacitated in ways beneficial to the practice of Euro-American science and its commensurability with capitalist logics.

Stem cell science, then, effects a symbolic reduction wherein bodies are reduced to raw material, value is reduced to the cognitive labor of the scientist, and stem cells (and the ova from which they are derived) are reduced to capital. Waldby and Cooper make plain that female/feminized bodies become a resource that can be extracted from areas marked by extreme precarity, in part because their biology becomes an asset for others, not a property of the person. The kind of fragmentation of the body that stem cell science affords is not only the result of advances in biological research and its increased specialization, but also of the need for capital to find profit in the body—as commodity, as source of speculative value, and as promised future profits. The body ("life itself") is synecdochically reduced to value in the narrow, capitalistic sense of the term.

Conclusion

The theory of coproduction holds that science, and the social relations of which it is a part, are both analytically distinct (i.e., we can study the practices of scientists without necessarily talking about the larger culture outside of the laboratory) and inextricably linked. For those scholars interested specifically in the coproduction of science and capital, this ambivalence means that although we can talk about, say, the different meanings of innovation and speculation obtained in the distinct spheres of science and capital, these nevertheless are tropes that enable the restructuring of relations and practices in both spheres according to market logics. And as Lukas Rieppel, Eugenia Lean, and William Deringer argue, coproduction has a long and global history in which "epistemic and commercial values—matters of fact and matters of exchange" were co-articulated well before the 19th century, the time frame during which what we consider "modern" science and capitalism arose and became objects of study.[56]

Coproduction does not, however, quite capture the ways in which the global circulation of both Euro-American science and capital, and their respective albeit intersecting logics, have produced the synecdochal reduction of all science to Euro-American science. This symbolic reduction has been enabled by the expropriation of land, labor, and life that is not only characteristic of the history of global capitalism and its various stages but has made possible the practice and export of Euro-American science, from agriculture to the body itself, as a source of value for biocapitalism.

Nevertheless, the notion of coproduction does enable scholarly engagement with the unexpected outcomes of the global circulation and engagement of Euro-American science/capital with localized histories and practice, or what Rieppel, Lean, and Deringer describe as capital and science's "complicated circuits, unanticipated trajectories, and feedback loops."[57] Thus, through a coproduction lens, the critic can look not only at the emergence of counter-ideology[58] but also at how new knowledges, relations, and subjectivities emerge from the circulations of science/capitalism. Along these lines, Lewontin posits that farmers in the developing world become a proletariat as a result of biotech's transformation of living matter,[59] and Waldby and Cooper draw on feminist labor theory to propose that ova vendors might identify not as donors or research subjects but as workers entitled to a say in whether, and how, they participate in the global bioeconomy.[60] In neither case is the goal to return to conditions prior to the bioeconomy's rise (as if such a possibility even existed) but rather to think through the agentic power of resultant positionalities and the political imaginaries that inhere within. By globalizing rhetorical studies of science, the goal is not to recuperate a lost past but to map the social relations that are formed as science and capitalism coproduce one another in temporally and spatially diverse ways.

Notes

1. Stephen Buranyi, "Big Pharma is Fooling Us," *New York Times*, December 17, 2020, https://www.nytimes.com/2020/12/17/opinion/covid-vaccine-big-pharma.html.

2. Matt Apuzzo and Selam Gebrekidan, "For Covid-19 Vaccines, Some Are Too Rich—And Too Poor," *New York Times*, December 28, 2020, https://www.nytimes.com/2020/12/28/world/africa/covid-19-vaccines-south-africa.html.

3. Richard Levins, "How Cuba is Going Ecological," in *Biology Under the Influence: Dialectical Essays on Ecology, Agriculture, and Health*, ed. Richard Lewontin and Richard Levins (New York: Monthly Review Press, 2005): 343–64.

4. Kaushik Sunder Rajan, *Biocapital: The Constitution of Postgenomic Life* (Durham, NC: Duke University Press, 2006). Sunder Rajan theorizes a Marxist approach to *coproduction*, the term coined by Sheila Jasanoff to capture the ways in which scientific and other cultural practices interanimate one another. See also Kaushik Sunder Rajan, "Introduction," in *Lively Capital*, ed. Kaushik Sunder Rajan (Durham, NC: Duke University Press, 2012).

5. For clarity, we employ the term *expropriation* to describe capital accumulation by (often violent) means other than wages and labor contracts. Expropriation, as a concept, helps elucidate the ways in which capitalism benefits from dispossession of land and resources, dependent or otherwise unfree labor (e.g., in private/domestic spaces, prison, and agriculture), the punitive effects of debt (debt leading to new kinds of deeply exploitative, low-wage work), and, as we'll see in the case of biocapitalism, the capture of the biological capacities of living matter as well as the creative energies of non-Euro-American science. We also assume,

with Nancy Fraser, that expropriation and exploitation often go hand in hand since the former provides the conditions for the latter. We develop this point in the last section of the chapter. See Nancy Fraser, "Expropriation and Exploitation in Racialized Capitalism: A Reply to Michael Dawson," *Critical Historical Studies* 3, no. 1(Spring 2016): 163–78.

6. Vandana Shiva, *Biopiracy: The Plunder of Nature and Knowledge* (Berkeley, CA: North Atlantic Press, 2016); Richard Levins and Richard C. Lewontin, eds., *Biology Under the Influence: Dialectical Essays on Ecology, Agriculture, and Health* (New York: Monthly Review Press, 2005).

7. Sunder Rajan, *Biocapital*. Genetic tests are an example of what he means by biocapital. Such tests are services that, when consumed, provide revenue for genomics companies. But there is a speculative finance aspect to this as well: when the sequences produced as a result of those tests are stored, they can be used to generate investment capital in the hopes that sequence data can be used to develop genome-based pharmaceuticals.

8. Thomas Kuhn, *Structure of Scientific Revolutions* (Chicago: The University of Chicago Press, 2012).

9. See his book *Biocapital* for an elaboration of a Marxist understanding of coproduction.

10. The merchant phase of capitalism, what Cook here aptly describes as investment-oriented, will figure into Sunder Rajan's theory of biocapitalism. Harold Cook, "Sciences and Economies in the Scientific Revolution: Concepts, Materials, and Commensurable Fragments," *OSIRIS* 33 (2018): 25–44.

11. Cook, "Sciences and Economies," 43–44. In some ways, Cook's observation is similar to Levins and Lewontin insofar as he identifies the formative role played by what we could call class values (in this case merchant values, a precursor to the bourgeoisie). For an elaboration of how statistics is an example of such commensurable knowledge, see Arunabh Ghosh, "Lies, Damned Lies, and (Bourgeois) Statistics: Ascertaining Social Fact in Midcentury China and the Soviet Union," *OSIRIS* 33 (2018): 149–68.

12. Richard Levins and Richard Lewontin, *The Dialectical Biologist* (Cambridge, MA: Harvard University Press, 1985), 268. For a feminist critique of Cartesianism see Evelyn Fox Keller, *Reflections on Gender and Science* (New Haven, CT: Yale University Press, 1985). For feminists like Keller, Cartesianism is a kind of masculinist perspective on the world, one marked by binaries, hierarchies, and control. Sandra Harding has written extensively on the ways in which such perspectives and methods also make science decidedly Euro-American. See for example Sandra Harding, ed., *The "Racial" Economy of Science: Toward a Democratic Future* (Bloomington: Indiana University Press, 1993).

13. Levins and Lewontin, *Dialectical Biologist*, 269. For example, parts "are ontologically prior to the whole" and are "homogenous within themselves"; "causes

are separate from effects, causes being the properties of subjects, and effects the properties of objects."

14. Levins and Lewontin, *Dialectical Biologist*, 269–70.

15. Levins and Lewontin, *Dialectical Biologist*, chapter 10.

16. Levins and Lewontin, *Dialectical Biologist*; Lewontin and Levins, *Biology Under the Influence*.

17. Edward Yoxen, "Life as a Productive Force: Capitalising the Science and Technology of Molecular Biology," in *Science, Technology, and the Labour Process: Marxist Studies Volume 1*, ed. Les Levidow and Bob Young (London: CSE Books, 1981), 66–122.

18. Yoxen, "Life as Productive Force," 102; See also Lily Kay, *The Molecular Vision of Life: Caltech, The Rockefeller Foundation, and the Rise of the New Biology* (New York: Oxford University Press, 1993). Situating such influential actors as applied industrial research, federal funding of research, and the Rockefeller Foundation (whose president, Warren Weaver, coined the term *molecular biology* in 1938) as central to the history of the emergence of molecular biology (as distinct from classical genetics or biochemistry), Yoxen argues that the restructuring of labor in the sciences—the "re-skilling or the planned reconstitution of biological labour power, in increasingly specialised forms"—was the effect of varied material forces from the late 19th century to postwar state interventions aiming for a managerial approach to scientific research.

19. Yoxen, "Life as Productive Force," 112.

20. Yoxen, "Life as Productive Force," 74. He is influenced here by cultural studies and the work of Raymond Williams and Stuart Hall, among others.

21. Shiva demystifies the language of "banking," calling such practices "biopiracy." See Shiva, *Biopiracy*.

22. Yoxen, "Life as Productive Force," 112. We understand his use of the word *appropriation* to mean *expropriation*. Fraser (2016) argues that these terms tend to be used interchangeably along with *dispossession*. For Lewontin's argument, see "The Maturing of Capitalist Agriculture: Farmer as Proletarian," in *Biology Under the Influence: Dialectical Essays on Ecology, Agriculture, and Health*, ed. Richard Lewontin and Richard Levins, 329–41. As Sunder Rajan explains, the development of recombinant DNA technology in the 1970s allowed the life sciences to become "technological" and "industrial" insofar as it enabled scientists to engineer new configurations of DNA and in corporate contexts (*Biocapital*, 5). Once the biotech industry got going, more resources were directed toward the science of genetic engineering. Sunder Rajan notes a number of developments that enabled this coproduction, including: private finance and public research dollars (e.g., venture capital and NIH funding); policy developments that helped forge connections between academia and the private sector (e.g., the Bayh-Dole Act); and a supportive legal climate (e.g., gene patenting, which shows how tech-

nological developments in gene sequencing informed ongoing reconfiguring of patent law, while the actual production and use of gene sequences was informed by their ever-evolving legal status). Sunder Rajan, *Biocapital*.

23. Kean Birch and David Tyfield, "Theorizing the Bioeconomy: Biovalue, Biocapital, Bioeconomics or . . . What?," *Science, Technology, and Human Values* 38, no. 3 (2012): 322.

24. Birch and Tyfield, "Theorizing the Bioeconomy," 317–18. They write that property rights provide the incentive (and overcome obstacles to) "the appropriation of knowledge (e.g., patents), nature (e.g., biomass), and biological matter (e.g., vital fragments) . . . knowledge, specifically, has to be made a scarce asset (i.e., a monopoly) in order to capture value, which means that IPRs are central to the extraction of rent from diverse forms of knowledge labor."

25. Birch and Tyfield, "Theorizing the Bioeconomy," 317. We take *appropriation* here as an example of what we are calling *expropriation*.

26. As Lukas Rieppel, Eugenia Lean, and William Deringer argue in the introduction to a special issue of *OSIRIS* on novel approaches to the histories of science and capitalism, what is considered "knowledge" is the product of circulation, not its driver, and, moreover, circulation has never *not* been transnational in scope. Furthermore, approaches that view modern science as the product of capitalism (as capitalism spreads, so does science) or that view advancements in science and technology as what enabled capitalism to develop in Europe (the so-called "divergence" theory) are untenable. See Lukas Rieppel, Eugenia Lean, and William Deringer, "Introduction: The Entangled Histories of Science and Capitalism," *OSIRIS* 33 (2018): 1–24.

27. Indigenous environmental scientist Jessica Hernandez argues that we should use the world *science* not *knowledge* to denote non-Euro-American epistemologies and methodologies. As she puts it, "I use a persona of an Indigenous scientist . . . because I use the Western sciences, the training that I have in the physical and environmental sciences to advocate for the inclusion of Indigenous ways of knowing that I refer to as Indigenous science. And I think that oftentimes the term that we use a lot is traditional ecological knowledge. But when I have seen that being introduced into the environmental sciences, a lot of scientists kind of focus more on the traditional and they continue to speak about Indigenous peoples, our ways of knowing in the past tense. And the reason why I use the word science is because our knowledge has adapted. We have survived colonization. We're still surviving climate change impacts. As we know, right, climate change is already impacting our Indigenous communities. And I see it as a science. Especially the way that science is formulated, we are still making questions, we're still making observations. It's just that the methods or the ways that we passed on our knowledge is very different than it's done in Western science where you publish peer-review articles, where you collect numerical data. And

I think that is different but in the same way it's kind of still ongoing knowledge that adapts and formulates new knowledge as we speak today." See her interview on the radio show *Science Friday*, April 22, 2022, https://www.sciencefriday.com/segments/indigenous-science-climate-change/#segment-transcript.

28. Sunder Rajan, *Biocapital*, 14. Italics in original.

29. Sunder Rajan, *Biocapital*, 14.

30. Sunder Rajan, "Introduction," 7–8.

31. Sunder Rajan, *Biocapital*, especially Chapter 5.

32. Richard Levins, "Science and Progress: Seven Developmentalist Myths in Agriculture," in *Biology Under the Influence: Dialectical Essays on Ecology, Agriculture, and Health*, ed. Richard Lewontin and Richard Levins (New York: Monthly Review Press, 2005), 321.

33. Kalindi Vora, *Life Support: Biocapital and the New History of Outsourced Labor* (Minneapolis: University of Minnesota Press, 2015).

34. Yoxen, "Life as a Productive Force," 110.

35. Stefan Helmreich, "Blue-Green Capital, Biotechnological Circulation and an Oceanic Imaginary: A Critique of Biopolitical Economy," *BioSocieties*, 2 (2007): 293.

36. Helmreich, "Blue-Green Capital," 294. Here he uses the term *reproductive* in line with other scholars who adapt the Marxist notion of productive labor to theorize the capturing of vitality by the biological sciences.

37. Donald Lowe, *The Body in Late Capitalist USA* (Durham, NC: Duke University Press, 1995).

38. Vora, *Life Support*, 3.

39. Vora, *Life Support*, 3.

40. Krista Conger, "Researchers Awarded $31 Million for Clinical Trials to Treat Stroke, Heart Failure, Brain Cancer," Stanford Medicine News Center, *Stanford Medicine*, September 9, 2021, https://med.stanford.edu/news/all-news/2021/09/grants-for-stem-cell-clinical-trials.html.

41. According to the NIH, the only approved therapy is for blood disorders.

42. Catherine Waldby et al., "From Altruism to Monetisation: Australian Women's Ideas About Money, Ethics, and Research Eggs," *Social Science & Medicine*, 94 (2013): 35. Research has shown that people are more likely to donate IVF-related embryos than ova, although not at rates that researchers would like.

43. Embryos are not necessarily needed for stem cell research, as oocytes alone can further the goals of regenerative medicine. In the United States, it has been the norm that ova donation for fertility can receive some compensation (e.g., for travel), but not for research. That is changing as professional guidelines are revised and states begin permitting payment for stem cell-research related ova donation. For research (based on stem cell research in New York state) on how paying for egg production and extraction increases supply considerably, see L. Zakarin Safier et al., "Compensating Human Subjects Providing Oocytes for Stem

Cell Research: 9-Year Experience and Outcomes," *Journal of Assisted Reproduction and Genetics* 35 (2018): 1219–25.

44. Catherine Wallaby, *The Oocyte Economy: The Changing Meaning of Human Eggs* (Durham, NC: Duke University Press, 2019). We should note that scientists receiving federal money are not permitted to create embryos for their research.

45. Catherine Waldby and Melinda Cooper, "The Biopolitics of Reproduction: Post-Fordist Biotechnology and Women's Clinical Labor," *Australian Feminist Studies* 23, no. 55 (2008): 63.

46. Waldby and Cooper, "Biopolitics of Reproduction," 64.

47. While Birch and Tyfield, "Theorizing the Bioeconomy," focuses on immaterial labor, Waldby and Cooper demonstrate the critical if underacknowledged role that material (including reproductive) labor plays in value-making. See Waldby and Cooper, "Biopolitics of Reproduction." An even earlier feminist theorization of labor can be found in Carolyn Merchant, *The Death of Nature: Women, Ecology, and the Scientific Revolution* (New York: HarperCollins, 1980), in which she argues that early scientific ideas embracing domination of "man" over (gendered female) "nature" were intelligible and persuasive given the dominance of capital over labor more broadly.

48. Waldby and Cooper, "Biopolitics of Reproduction."

49. Waldby and Cooper, "Biopolitics of Reproduction," 59.

50. Waldby and Cooper, "Biopolitics of Reproduction."

51. Waldby and Cooper, "From Reproductive Work to Regenerative Labour: The Female Body and the Stem Cell Industries," *Feminist Theory* 11, no. 3 (2010): 8–9.

52. Waldby and Cooper, "From Reproductive Work to Regenerative Labour," 8.

53. Waldby and Cooper write that similar to how "the generosity of the other woman is presumed" in transactional relations in biotech, "there is a similar tendency in feminist bioethical work to want to resolve the power relations by prohibiting commodification in favor of the gift from one woman to another. This ethics of generosity ends up institutionalizing the self-sacrifice of the other woman . . . and as analysis of the human tissue gift economy demonstrates, gifting under contemporary conditions of highly capitalized life sciences is often simply a way to expropriate donors and deny them rights over their bodily material." "Biopolitics of Reproduction," 67. We would add that this is another synecdochal reduction: that all ova-bearing persons are presumed to live under the label "woman," which then accounts for the discourses of altruism and "gift."

54. Waldby and Cooper, "From Reproductive Work to Regenerative Labour," 15.

55. Waldby and Cooper, "From Reproductive Work to Regenerative Labour," 17.

56. Rieppel, Lean, and Deringer, "Introduction," 9.

57. Rieppel, Lean, and Deringer, 13.

58. For example, the work of Levins and Lewontin as both activist scientists and public intellectuals enlightens the public about the need for skepticism

of science's truth claims and the need to rethink the world in which scientific research is carried out. If biology can be ideological, then competing ideological conceptions must be articulated and circulated in the public sphere, and these competing visions must be articulated to ongoing anti-capitalist movement struggles for economic and social justice both in the United States and around the world.

59. Richard Lewontin, "The Maturing of Capitalist Agriculture: Farmer as Proletarian," in *Biology Under the Influence: Dialectical Essays on Ecology, Agriculture, and Health*, ed. Richard Lewontin and Richard Levins (New York: Monthly Review Press, 2007), 329–42.

60. Waldby and Cooper, "Biopolitics of Reproduction"; "From Reproductive Work to Regenerative Labour."

Chapter Two

The Shifting Rhetoric of Environmental Science in Australia

Acknowledging First Nations People and Country

EMILIE ENS, SHAINA RUSSELL, BRIDGET CAMPBELL,
SABINA RYSNIK-STECK, MONICA FAHEY, RENEE CAWTHORNE,
PATRICK COOKE, AND DANIEL SLOANE

Pre-colonization, Colonization, and Colonial Rhetoric

The first British settlers described Australia as *terra nullius*, which was the first act of colonial rhetoric inflicted on First Nations Australians. Ironically, however, that "empty" land is now speaking back—through wildfires, drought, and other ravages of climate change—in ways that are catalyzing a rebalancing of power between First Nations peoples and Euro-American scientists. Through two case studies of environmental decision-making and fire management, this chapter describes how, since colonization, rhetorics of Euro-American science have severely impacted Australian First Nations peoples' ways of life and environmental management practices. We also, however, draw attention to the ways in which Euro-American scientists have begun to acknowledge the efficacy of First Nations environmental management practices and are seeking to learn from them to cope with climate change. Despite this promising course reversal, we conclude that significant advances in devolution of power, understanding of First Nations knowledge systems, and supporting First Nations-led initiatives are required to maintain a trajectory of reconciliation in Australian conservation. Therefore, this chapter takes an ironic approach to global rhetorics of science by revealing the injustices that have resulted from the colonial imposition of Euro-American science in First Nations communities, alongside the ways in which First Nations rhetorics of science have resisted colonization by Euro-American science. In short, our case studies show that Euro-American science, when practiced in contemporary Australia in collaboration with First Nations communities, is actually becoming a hybrid of Euro-American and First Nations practices.

First Nations Australians have managed the land, freshwater, and seascapes of the Australian continent since time immemorial (c. 65,000 years in some areas, according to Euro-American scientific estimates).[1] First Nations environmental "management," commonly referred to as "caring-for-Country" in contemporary Australia, is interconnected with other aspects of the distinct and diverse lifeways of First Nations societies and governed through complex sociocultural systems of kinship and land tenure. While caring-for-Country practices continue or are being revitalized in some areas across much of the continent, they were largely disrupted, discouraged, and forcibly prohibited following British colonization in 1788. From the time of colonization, British land management practices were transplanted onto the Australia continent. The introduction of foreign livestock (hard-hoofed herbivores), crops, and pests resulted in the destruction of many First Nations–managed ecosystems both intentionally in fertile farming areas and incidentally as some of these species "jumped the fence" and became invasive in nonagricultural settings.[2] First Nations Australians have fought to achieve recognition as the Traditional Owners and managers of Australia for more than 230 years, a journey articulated largely through the land rights movement (see below). As reflected in national policy, the rhetoric surrounding First Nations Australians has changed dramatically over time, with greater articulation of First Nations perspectives and realities in contexts such as land management and science.

At the time of colonization, between 300,000 and 1 million First Nations people were organized into over 750 language groups and corresponding nations.[3] The clan or tribal estate is the primary unit of First Nations Australian land tenure, and socio-ecological management is governed through complex systems of kinship and ceremonial obligations.[4] First Nation estate boundaries were largely invisible to colonizers, for whom the culturally accepted system of land tenure manifested in fences, walls, and written title deeds.[5] The different modes of land management were also largely invisible, and the "lifeways" of First Nations Australians was subsumed under racist rhetorics of the "primitive nomadic hunter gatherer."[6] The perpetuation of this rhetoric continued despite evidence of large-scale sedentary agricultural and fisheries operations, long-term food storage infrastructures, and the (also seemingly invisible) sustainable management of complex biodiverse systems across the entirety of the continent.[7] This colonial rhetoric surrounding First Nations Australians worked hand in hand with the concept of *terra nullius* (land belonging to no one) and the Eurocentric social Darwinist rhetoric that facilitated and culturally validated the invasion and dispossession of First Nations peoples' lands to make way for the encroachment of the British empire.[8] This pressure forced many First Nations Australians, who were also impacted by the introduction of foreign diseases (smallpox, influenza, and measles), into inland territories of other clans, or into frontier warfare.[9] However, it is important to note that different groups were and still are impacted differently by colonization, with those from the southeastern coastal regions at the forefront of the invasion compared to those in remote desert communities who continued traditional lifestyles until the mid-1980s.[10]

The Evolution of Colonial Rhetoric and Policy

Colonial policies that were introduced to control, restrict, and define First Nations peoples lives over the past two centuries reflect the changing rhetoric surrounding First Nations Australians. Initial policies of protection and segregation from the late 19th to mid-20th centuries perpetuated the hegemony of British superiority. First Nations knowledge, laws, customs, and modes of social organization were disregarded as "primitive" and "heathen" and efforts were made to deny First Nations people of any agency in order to "protect" them from their own culture and way of life.[11] During this time, the forced segregation of First Nations people in Christian missions was a key strategy used to dispossess, disrupt, and deny the continuation of First Nations culture and language.[12] The following policy of assimilation led to further atrocities, including the "Stolen Generations."[13] Here the underlying colonial rhetoric remained, driving and justifying actions to integrate First Nations Australians considered not of "full blood" into the wider population by taking them away from their families and enculturating them into white society.[14]

Change arrived after the 1967 referendum, which granted citizenship to First Nations Australians. Policies of integration and self-determination were introduced, following the changing rhetoric that finally granted First Nations Australians rights (for example, voting and equal pay) that allowed them to regain control of their own lives and be part of an Australian society that had increased in diversity following the abolishment of the white Australia policy.[15] In northern Australia this rights movement sparked the outstation movement, where clan groups were encouraged and funded to return to Country and establish "homelands" or "outstations."[16] The era of reconciliation began in 1991, led by rhetoric that acknowledged the misdeeds of the past and stated that greater rights, respect, and communication were needed to ensure a better future. The changing rhetoric that enabled the establishment of these later policies did not arise within the Australian government independent of First Nations voices and action. Rather, the transformation of rhetoric and policy was catalyzed by First Nations peoples efforts to gain recognition and rights, largely articulated through the land rights movement.

The Land Rights Movement:
Recognizing First Nations Environmental Management and Science

The land rights movement was a major catalyst of change in Australian environmental management decision-making and control from the ground up. A succession of state and territory land rights acts started in 1976 in South Australia and the Northern Territory following ongoing community pressure for recognition of ancestral clan estates.[17] This articulation of First Nations voices and agendas and affirmation of agency over land and sea saw the translation of Aboriginal cultural Law/Lore into mainstream Australian law. This process was epitomized by the decade-long legal struggle of Eddie Mabo to gain recognition of his ancestral estate in the Torres Strait, which resulted in the national Native Title Act 1993.[18] These decades of ongoing struggles, still occurring on many ancestral estates, demonstrate how rhetoric was and still is being used to define and control First Nations peoples' lives. Not only was the Euro-American hegemonic rhetoric pervasive with regard to rights to access and decision-making about environments, but it also determined how these environments were understood and analyzed through Euro-American science.

Euro-American science often describes the environment in terms of productivity and diversity—in other words, whether or not it has value to settlers. This rubric determined how the land was managed and who could "own" it and make decisions about it. For example, areas of high soil nutrients and rainfall were identified and forcibly possessed for agricultural areas, literally resulting in the clearing of previous First Nations land management and any associated cultural history. First Nations voices were and still are largely ignored in decisions about land use at the larger scale. Similarly, the rhetoric of Euro-American science has been used to justify colonial approaches to fire management across the vast continent of Australia since colonization, largely void of First Nations expertise until recently. These examples of the shifting power in environmental management and science, as a result of the pressure of both First Nations and their supporters, are explored in more detail below.

ADVOCATING FOR THE INCLUSION OF FIRST NATIONS PEOPLES AND KNOWLEDGE IN ENVIRONMENTAL MANAGEMENT

The shift in environmental decision-making towards inclusion of First Nations perspectives since the mid-20th century was complemented by international pressures, such as those from the United Nations, to address social inequality and inclusion of First Nations and local peoples in all relevant forums, including environmental conservation. Examples of such international directives come from the UN's Convention on Biological Diversity and Declaration on the Rights of First Nations Peoples. Therefore, bottom-up and top-down pressures drove more inclusive environmental legislation and decision-making in Australia, culminating in the Indigenous Protected Area program (1997), the Working on Country (2007) ("Ranger") program and the Australian government's Strategy for Nature (2019–2030). The international and national mandates for inclusion of First Nations peoples and knowledges have opened up the discourse on the rhetoric of science to be more inclusive, not just of First Nations and local peoples but also their knowledge systems. The decolonization and decentralization of environmental decision-making, practice, and science is increasingly being advocated for in an effort to dismantle Euro-American scientific hegemony and balance power relations between First Nations peoples and settlers with the goal of improved environmental management outcomes at local to regional scales.[19]

The Tensions between Euro-American Science and First Nations Knowledge

Euro-American science can be defined as a system of knowledge based on scientific laws and influenced by the hypothetico-deductive positivist scientific method. In order to yield unbiased results, this form of science pressures the scientist to be objective. This assumption of distance between subject and scientist has often resulted in the view that Euro-American culture is separate and above the natural world.[20] Euro-American science has predominantly shown a callous disregard towards knowledges from other cultures,[21] which has led to a narrow worldview and lack of deeper understanding, particularly considering its brief history in comparison to much older and developed First Nations Knowledge systems. However, this lack of understanding is starting to be revealed.[22] The recognition of the importance and integration of First Nations Knowledge, particularly in relation to socio-ecological sustainability (with the long term multigenerational information collected by First Nations cultures), is now slowly being understood and accepted by the Western canon.[23]

First Nations Knowledge is the accumulated intergenerational knowledge of First Nations peoples that has been gained through a long-term connection with place and is usually transferred orally. Compared to Euro-American science, it is an all-encompassing, holistic knowledge system in which all aspects of life are intertwined and connected.[24] It is knowledge gained through multilevel understandings, both physical and spiritual, that have allowed people to survive, thrive, and understand the world around them and their place within it.[25] Euro-American science, on the other hand, has sought to separate areas of study into different research disciplines, at least until the recent rise of multidisciplinary research. Nevertheless, both knowledge systems share the core aim of understanding the world around them.[26]

Kinship and First Nations Australian Environmental Management

In the Australian context, kinship is at the heart of First Nations Knowledge, social relations, and environmental management. The First Nations Australian kinship system is complex. It establishes an individual's relationship to others and the biotic and abiotic world around them, as well as dictating roles, responsibilities, and obligations within the community.[27] The kinship system is classificatory and includes hierarchical levels including moieties, sections, subsections, and clans, which can all have different totems associated with them.[28] All biotic and abiotic environmental components are encompassed in the moiety system, which is centered on the understanding that everything is split in half, and each half is a mirror of the other. In order to fully understand the world, these two halves must come together. The kinship system also determines social relationships such as marriage, and who is permitted to be in the presence of whom (e.g., avoidance relationships), which can have implications in contemporary environmental management activities. All totems are preordained with the exception of the personal, which identifies that person's individual spirit. The individual or group does not own totems, but rather they become custodians of them. The totemic system allows connections to be made between the individual and the physical world and dictates protection and exploitation rights.[29]

The significance of kinship in First Nations environmental ethics has been referred to as an First Nations ethic of connection[30] and kincentric ecology.[31] The reciprocal human–Country relationship is exemplified by First Nations fire management, currently referred to as cultural burning, which is characterized by low intensity and patch burning that draws on knowledge of weather patterns and vegetation to "keep the Country clean" and accessible by preventing a buildup of understory biomass.[32] When First Nations Australians burn Country according to First Nations Law/Lore, the canopy is respected and must not burn.[32] The canopy provides habitat, protection, shade, and water for birds, arboreal mammals, and ground-dwelling mammals. Cultural burning is about respecting the canopy and those animals that depend on it. It is also about protecting totems, and adhering to one's cultural responsibilities through kinship. This holistic ethic of care is the sociocultural fabric that sees humans and Country in a reciprocal contract with associated responsibilities, rights, and obligations.

Examples of the Changing Rhetoric in Environmental Management of Australia

Example 1: Australian Indigenous Protected Area and Indigenous Ranger Programs

In the 1980s, government initiatives including the Community Development Employment Projects (CDEP) were developed to offer economic support for

Indigenous rangers in Australia's conservation agenda.[33] While the CDEP was driven largely by socioeconomic (employment) objectives, in 1997, the Indigenous Protected Area (IPA) program followed, which provided operational funding for Indigenous rangers and community members to manage declared areas of their Traditional estates. The IPA program served a conservation agenda and contributed to Australia's National Reserve System (NRS) which was designed to conserve comprehensive, adequate, and representative parcels of Australia's biogeographical regions. The IPA program is premised on the goodwill of Traditional Owners volunteering their land as part of Australia's conservation system. Once the Australian government accepts an IPA proposal, a consultation period begins during which the Traditional Owners, proposing organization, and government representatives negotiate terms in line with International Union for the Conservation of Nature (IUCN) categories V or VI. The flexibility in following IUCN category V or VI is welcomed by Traditional Owners as many aspire to manage their estates not only for natural and cultural conservation but also for socioeconomic development. In 2020, there were seventy-five IPAs accounting for more than 44 percent of Australia's Natoinal Reserve System. While it started as a conservation agenda, the rhetoric surrounding the IPA movement promotes an approach to conservation that aims to enhance First Nations control in management and recognize the cultural, spiritual, and economic significance of land for the improved economic development of First Nations peoples through the paradigm of caring-for-Country.[34] This conservation-as-development model encompasses two distinct value systems operating in the intercultural space of First Nations natural and cultural resource management: local practice/customary management and bureaucratic requirement.[35] Often these values are contested, and power struggles occur over ideas of good governance.[36]

The shift from socioeconomic to conservation objectives is reflected in the discourse, from the First Nations concept caring-for-Country to government initiatives titled "working on country" and "caring for our country." This change in discourse correlates with a shift in funding for ranger groups, based on social development models that afforded local cultural values (at least in theory) equal footing to Euro-American conservation values.[37] This shift occurred in the context of expansion of the Australian government's Indigenous development policies that resulted in programs such as The Intervention, Closing the Gap, Stronger Futures, and the Indigenous Advancement Strategy. Many of these programs have been heavily criticized by some groups of First Nations Peoples and non-First Nations advocates as they focus on the deficit issues of poor health, education, employment, and violence rather than taking the strengths-based, First Nations-led approach that has been used in other parts of the world, such as Canada. Much of this deficit rhetoric has pervaded the First Nations environmental management and governance discourse (and resulting funding structures) where the lack of administrative and computer skills, for example, create a need to employ non-First Nations staff. The failure to recognize and appreciate First Nations conceptions of work and culturally informed workplace governance and administration is an ongoing challenge in Australian First Nations organizations that rely on Euro-American funding models.

Nevertheless, the shifts in the focus of Australian government funding from community development and social justice towards a natural and cultural resource management focus through the IPA and ranger programs have enabled First Nations Australians to find innovative ways to articulate their cultural and spiritual values and assert their rights as the decision makers and drivers of conservation. In contemporary First Nations land and sea management, First Nations groups draw on blends of traditional and Euro-American science and innovation depending on the task and skills at hand.[38] These innovations include adoption of customary kinship Laws in governance, community planning, and prioritization to assert connections between people, places, plants, and animals. For example, when conducting cultural burns, many First Nations groups ensure that the correct Traditional Owners or Traditional Managers of Country are supportive of the burn and physically initiate the burn. The reinvigoration of cultural Laws into practice also informs values and issues at the landscape scale and guides the use of relevant communication tools with a focus on visual, non-text formats, such as oral communication (having a yarn), workshops, and video.[39] The Australian First Nations conservation movement has reasserted clan-based relative autonomy, resistance, and articulation and is regarded as a propitious niche where First Nations and Western conservation aspirations have come together for a common goal and for mutual benefit.[40] Notwithstanding, we acknowledge that there are ongoing challenges to cross-cultural conservation that require constant renegotiation of aspirations, priorities, and actions to facilitate mutual benefits.

Global and national policies continue to mandate the need for partnership approaches and recognition of First Nations peoples and their knowledges in the protection and management of natural environments.[41] Mirroring these developments, the focus of Australian conservation and environmental management programs continues to expand to include First Nations values and perspectives; however, Traditional Owners operating within the IPA system are often frustrated with the lack of recognition for their environmental management activities, despite the government rhetoric.[42] For example, although the propitious niche for the collaboration and mutual benefit of First Nations and non-First Nations stakeholders is recognized, Euro-American worldviews and priorities continue to dominate decision-making, especially when time and money are limited.[43] Nevertheless, there has been an obvious shift in control of the conservation estate, with IPAs making up a growing and substantial proportion of the national conservation agenda and land still being "handed back" to Traditional Owners through the legal system. This is in contrast to other invaded First Nations states where Traditional First Nations land ownership is still contested and denied, such as in many parts of Asia.[44]

Example 2: Fire Management

In Australia, like other parts of the world, First Nations Peoples have made the greatest headway in recent challenges to the dominant Euro-American rhetoric of science in the area of fire and landscape burning. Prior to European colonization of Australia starting in 1788, Australian First Nations groups developed detailed knowledge and use of fire from small to large scales over millennia.[45] Fire is integral to First Nations people's culture and daily lives, for example in cooking, procuring food and other resources, communication, cleaning Country, and ceremony.[46] However, with European colonization came European control of landscapes, including the use of fire. European norms of keeping fire out of ecosystems and controlling fire became the dominant mantra of natural resource management with a strong focus on protecting infrastructure and farming lands. It was only in the last forty years or so that First Nations fire knowledge came back into mainstream Australian conservation discourse, particularly in remote northern and central Australia. This movement has now spread across the nation and is referred to as Right-Way Fire or cultural burning. The breakdown of Euro-American hegemony in fire science and the inclusion of First Nations people and knowledge was fueled by the growing Indigenous ranger movement,[47] the emergence of the carbon trading market,[48] First Nations human and land rights recognition, and the increase in First Nations control and decision-making over ancestral clan estates (as described above).[49] Innovation in First Nations burning and carbon abatement science in Australia has also become an inspiration for First Nations peoples in other colonized nations who similarly aspire to reinstate First Nations knowledge and burning practices.[50] However, despite advances in this space, significant barriers and challenges in reconciling the differences between First Nations and Euro-American ways of knowing and doing persist in fire management decision-making and practice.

Initial breakthroughs in cross-cultural fire science and management were documented in northern and central Australia.[51] Significant collaborations between Traditional Owners, First Nations community rangers, scientists, and fire agencies in northern Australia worked to build evidence of the environmental, social, and carbon abatement benefits of the patchy low-intensity burns of traditional fire management in Arnhem Land.[52] Documentation of First Nations burning practices and development of partnerships were simultaneously occurring in Cape York,[53] the central desert,[54] and the Kimberley region of northern Western Australia,[55] serving to build evidence for the benefits and processes involved in Aboriginal use and management of fire.

The large, sparsely populated areas of northern and central Australia have afforded Traditional Owners the opportunity to maintain and reinvigorate traditional fire management practices since the 1980s following the return to Country "outstation movement."[56] In northern and central Australia, Traditional Owners actively deployed rhetoric by asserting their cultural obligations to burn Country and demonstrated the low environmental impact of cultural burning. As a result, they developed crucial partnerships with government fire services and scientists that enabled the reactivation and documentation of the social and ecological benefits of cultural burning.[57] Although we acknowledge here that even the most remote places were influenced by European missions and settlements, Traditional Owners describe still being able to go out on weekends to hunt, gather, burn Country, and maintain traditional practices.[58]

The cross-cultural partnerships that developed in northern Australia around the reinvigoration of First Nations burning led to detailed evidence building of the multiple benefits of First Nations fire management, including carbon abatement, vegetation and fauna protection, and cultural, social, emotional, health, and economic benefits.[59] This mounting evidence was used to drive the large-scale implementation of traditional fire management practices, which were intertwined with use of modern technologies such as helicopters, incendiaries, and remote sensing.[60] Once these methods had been proven by Indigenous ranger groups, scientists, and community development activists to deliver beneficial outcomes, an expansion of cultural burning practice occurred across northern and central Australia and, later, in southeastern Australia where fire management was (and still is) tightly controlled by Euro-American science and ideology.[61] Nevertheless, there is significant cultural burning now occurring across southeastern Australia, which includes the decolonization of fire management practice and community education about the benefits of cultural burning.[62] The recent increased community uptake of cultural burning has been further fueled by the recent large-scale uncontrolled wildfires in populated parts of southeastern Australia, which have caused widespread damage to infrastructure, human livelihoods, biodiversity, and human lives.[63]

Conclusion

This chapter has demonstrated how taking the trope of irony as a guide to examining the public practice of science in Australia exposes key sites of conflict between Euro-American and First Nations practices, which can serve as occasions for reinventing a more productive and just dynamic. Environmental decision-making and fire management practices in Australia are embracing once-denounced First Nations knowledge systems, and the Euro-American-scientific hegemony and rhetoric of colonial Australia is dissolving. However, these gains have been hard fought over centuries, and there is still a long way to go until there is meaningful and equitable recognition and inclusion of First Nations knowledge and people in Australian environmental science and management. Only recently has the precolonial history of Australia started to be taught in schools and entered mainstream dialogue, including some First Nations languages. The rhetoric of science obliterated First Nations customary knowledge in many parts of Australia. Now, First Nations descendants and their non-First Nations allies have the very difficult task of recovering knowledge and language that was not documented but transmitted orally through story, song, dance, and art. A substantial increase in funding and support for First Nations Australian languages and cultures is required to stem the tide of First Nations cultural and language loss, as the majority are considered extinct or highly endangered.

Herein lies an opportunity for environmental scientists to engage in research and conservation combining natural and cultural resource management, or biocultural conservation. The Indigenous Protected Area and ranger programs and the resurgence of First Nations cultural burning are demonstrating the way forward for practical reconciliation and the decolonizing of environmental science in Australia. However, we still have a long way to go and success will only come with an understanding of the ways in which scientific practices are sustained by public rhetorical practices. A willingness among all Australians is required to embrace diverse ways of knowing and doing, especially those of Australia's First Nations peoples who skillfully managed Australia's landscapes—underpinned by kinship and spiritual connections to all human and non-human entities—for thousands of generations.

Notes

1. Rupert Gerritson, *Australia and the Origins of Agriculture* (Oxford: Archaeopress, 2008); Bruce Pascoe, *Dark Emu: Black Seeds: Agriculture or Accident?* (Broome, Australia: Magabala Books, 2014); Chris Clarkson et al. "Human Occupation of Northern Australia by 65,000 Years Ago," *Nature* 547, no. 7663 (2017): 306–10, https://doi.org/10.1038/nature22968; see this source for the date of human settlement.

2. Pascoe, *Dark Emu*.

3. Doug Marmion, Kazuko Obata, and Jakelin Troy, *Community, Identity, Wellbeing: The Report of the Second National Indigenous Languages Survey* (Canberra: Australian Institute of Aboriginal and Torres Strait Islander Studies, 2014).

4. Ronald Murray Berndt and Catherine Helen Berndt, *The World of the First Australians* (Sydney: Lansdowne, 1981).

5. David Ritter, "The Rejection of Terra Nullius in Mabo: A Critical Analysis," *Sydney Law Review* 18, no. 1 (1996): 5; Bill Gammage, *The Biggest Estate on Earth: How Aborigines Made Australia* (Melbourne: Allen and Unwin, 2011) works as a counterpoint.

6. Val Attenbrow, *Sydney's Aboriginal Past: Investigating the Archaeological and Historical Records* (Sydney: University of New South Wales Press, 2010); Lynette Russell and Ian J. McNiven, "Monumental Colonialism: Megaliths and The Appropriation of Australia"s Aboriginal Past," *Journal of Material Culture* 3, no. 3 (1998): 283–99.

7. Gammage, *Biggest Estate on Earth*; Gerritson, *Australia and the Origins of Agriculture*; Pascoe, *Dark Emu*.

8. Stuart Banner, "Why Terra Nullius-Anthropology and Property Law in Early Australia," *Law & History Review* 23, no. 1 (2005): 95; Mark Francis, "Social Darwinism and the Construction of Institutionalised Racism in Australia," *Journal of Australian Studies* 20, no. 50–51 (1996): 90–105.

9. Sarah Colley and Val Attenbrow, "Does Technology Make a Difference? Aboriginal and Colonial Fishing in Port Jackson, New South Wales," *Archaeology in Oceania* 47, no. 2 (2012): 69–77; Bruce Elder, *Blood on the Wattle: Massacres and Maltreatment of Aboriginal Australians since 1788* (Sydney: New Holland Press, 2003).

10. Scott Cane, *Pila Nguru: The Spinifex People* (Sydney: Fremantle Press, 2002).

11. Attenbrow, *Sydney's Aboriginal Past*; Russell and McNiven, "Monumental Colonialism."

12. Tim Rowse, "Review: *The Contest for Aboriginal Souls: European Missionary Agendas in Australia*," *Aboriginal History* 42 (2018): 195–98.

13. Peter Read, *A Rape of the Soul So Profound: The Return of the Stolen Generation* (London: Routledge, 2020).

14. Karen Menzies, "Understanding the Australian Aboriginal Experience of Collective, Historical and Intergenerational Trauma," *International Social Work* 62, no. 6 (2019): 1522–34.

15. John Stone, "Fifty Years of Unremitting Failure: Aboriginal Policy Since the 1967 Referendum," *Quadrant* 61, no. 11 (2017): 62.

16. Herbert C. Coombs, Barrie G. Dexter, and Lester R. Hiatt, "The Outstation Movement in Aboriginal Australia," *Australian Institute of Aboriginal Studies Newsletter* 14 (1980): 16–23.

17. Elspeth A. Young, "Aboriginal Land Rights in Australia: Expectations, Achievements and Implications," *Applied Geography* 12, no. 2 (1992): 146–61, https://doi.org/https://doi.org/10.1016/0143-6228(92)90004-7.

18. Tim Rowse, "How We Got a Native Title Act," *The Australian Quarterly* 65, no. 4 (1993): 110–32.

19. Martin N. Nakata, *Disciplining the Savages, Savaging the Disciplines*. (Sydney:Aboriginal Studies Press, 2007); Linda Tuhiwai Smith, *Decolonizing Methodologies: Research and Indigenous Peoples* (London: Zed Books, 1999).

20. Kay Anderson and Colin Perrin, "'Removed from Nature': The Modern Idea of Human Exceptionality," *Environmental Humanities* 10, no. 2 (2018): 447–72.

21. Douglas L. Medin and Megan Bang, *Who's Asking?: Native Science, Western Science, and Science Education* (Cambridge, MA: MIT Press, 2014).

22. Mark Bonta et al., "Intentional Fire-Spreading by 'Firehawk' Raptors in Northern Australia." *Journal of Ethnobiology* 37, no. 4 (2017): 700–18; Gary D. Cook, Sue Jackson, and Richard J. Williams, "A Revolution in Northern Australian Fire Management: Recognition of Indigenous Knowledge, Practice and Management,"

in *Flammable Australia*, ed. Ross A. Bradstock, A. Malcolm Gill, and Richard J. Williams (Collingwood, Australia: CSIRO Publishing, 2012), 293–305.

23. Emilie J. Ens et al., "Indigenous Biocultural Knowledge in Ecosystem Science and Management: Review and Insight from Australia," *Biological Conservation* 181 (2015): 133–49, https://doi.org/ 10.1016/j.biocon.2014.11.008; Nathanael D. Wiseman and Douglas K. Bardsley, "Climate Change and Indigenous Natural Resource Management: A Review of Socio-Ecological Interactions in the Alinytjara Wilurara NRM Region," *Local Environment* 18, no. 9 (2013): 1024–45.

24. Ladislaus M. Semali and Joe L. Kincheloe, *What is Indigenous Knowledge?: Voices From the Academy* (London: Routledge, 2002).

25. Madhav Gadgil, Fikret Berkes, and Carl Folke, "Indigenous Knowledge for Biodiversity Conservation," *Ambio* 22 (1993): 151–56.

26. Arun Bala and George Gheverghese Joseph, "Indigenous Knowledge and Western Science: The Possibility of Dialogue," *Race & Class* 49, no. 1 (2007): 39–61; Jayalaxshmi Mistry and Andrea Berardi, "Bridging Indigenous and Scientific Knowledge," *Science* 352, no. 6291 (2016): 1274–75.

27. Berndt and Berndt, *The World of the First Australians*; Patrick McConvell, Piers Kelly, and Sebastien Lacrampe, *Skin, Kin and Clan: The Dynamics of Social Categories in Indigenous Australia* (Sydney: ANU Press, 2018).

28. Berndt and Berndt, *The World of the First Australians*, 231–38.

29. Berndt and Berndt, *The World of the First Australians*, 135.

30. Deborah Bird Rose, "Exploring an Aboriginal Land Ethic," *Meanjin* 47, no. 3 (1988): 378.

31. Enrique Salmón, "Kincentric Ecology: Indigenous Perceptions of the Human–Nature Relationship," *Ecological Applications* 10, no. 5 (2000): 1327–32.

32. Victor Steffensen, *Fire Country: How Indigenous Fire Management Could Help Save Australia* (Collingwood, Australia: CSIRO Publishing, 2020).

33. Jon C. Altman and Peter Whitehead, *Caring for Country and Sustainable Indigenous Development: Opportunities, Constraints and Innovation* (Canberra: Australian National University Press, 2003); Dermot Smyth, Peter Taylor, and Arthur Willis, eds., *Aboriginal Ranger Training and Employment in Australia, Proceedings of the First National Workshop* (Canberra: Australian National Parks and Wildlife Service, 1985).

34. Stan Stevens, "Indigenous Management," in *Conservation through Cultural Survival: Indigenous Peoples and Protected Areas*, ed. Stan Stevens (Washington: Island Press, 1997), 189–224.

35. Elodie Fache, "Caring for Country, a Form of Bureaucratic Participation. Conservation, Development, and Neoliberalism in Indigenous Australia," *Anthropological Forum* 24, no. 3 (2014): 267-286.

36. Frances Morphy, "Australia's Indigenous Protected Areas: Resistance, Articulation and Entanglement in the Context of Natural Resource Management,"

in *Entangled Territorialities: Negotiating Indigenous Lands in Australia and Canada*, ed. Françoise Dussart and Sylvie Poirier (Toronto, ON: University of Toronto Press, 2017), 70–90.

37. Sean Kerins, "The Future of Homelands/Outstations," *Dialogue* 29, no. 1 (2010): 52-60.

38. Erin L. Bohensky, James R. A. Butler, and Jocelyn Davies, "Integrating Indigenous Ecological Knowledge and Science in Natural Resource Management: Perspectives from Australia," *Ecology and Society* 18, no. 3 (2013), https://www.jstor.org/stable/26269334.

39. Bohensky, Butler, and Davies, "Integrating Indigenous Ecological Knowledge."

40. Morphy, "Australia's Indigenous Protected Areas"; Heather Moorcroft, "Paradigms, Paradoxes and a Propitious Niche: Conservation and Indigenous Social Justice Policy in Australia," *Local Environment* 21, no. 5 (2016): 591–614.

41. Emilie Ens et al., "Putting Indigenous Conservation Policy into Practice Delivers Biodiversity and Cultural Benefits." *Biodiversity and Conservation* 25, no. 14 (2016): 2889–906.

42. Bevlyne Sithole et al. *Aboriginal Land and Sea Management in the Top End: a Community Driven Evaluation* (Darwin, AUS: CSIRO Publishing, 2007).

43. Wayne Barbour and Christine Schlesinger, "Who's the Boss? Post-colonialism, Ecological Research and Conservation Management on Australian Indigenous Lands," *Ecological Management & Restoration* 13, no. 1 (2012): 36–41, https://doi.org/10.1111/j.1442-8903.2011.00632.x.

44. Emilie Ens et al., "Recognition of Indigenous Ecological Knowledge Systems in Conservation and Their Role to Narrow the Knowledge-Implementation Gap" in *Closing the Knowledge-Implementation Gap in Conservation Science*, ed. Catalina C. Ferreira and Cornelya F.C. Klütsch (Switzerland: Springer, 2021), 109–39.

45. David M. J. S. Bowman, "The Impact of Aboriginal landscape Burning on the Australian Biota," *New Phytologist* 140, no. 3 (1998): 385–410, https://doi.org/10.1111/j.1469-8137.1998.00289.x.

46. Steffenson, *Fire Country*.

47. Dermot Smyth, Peter Taylor, and Arthur Willis, *Aboriginal Ranger Training and Employment in Australia* (Canberra, Australia: Australian National Parks and Wildlife Service).

48. Jeremy Russell-Smith et al., "Improving Estimates of Savanna Burning Emissions for Greenhouse Accounting in Northern Australia: Limitations, Challenges, Applications," *International Journal of Wildland Fire* 18, no. 1 (2009): 1–18, https://doi.org/doi:10.1071/WF08009.

49. Marcia Langton, *Burning Questions: Emerging Environmental Issues for Indigenous Peoples in Northern Australia* (Darwin, Australia: Centre for Indigenous Natural and Cultural Resource Management, Northern Territory University, 1998).

50. Timothy Neale et al., "Walking Together: A Decolonising Experiment in Bushfire Management on Dja Dja Wurrung Country," *Cultural Geographies* 26, no. 3 (2019): 341–59.

51. Peter Kenneth Latz and G. F. Griffin, "Changes in Aboriginal Land Management in Relation to Fire and Food Plants in Central Australia" (Symposium on the Nutrition of Aborigines, Canberra, Australia, October 23, 1978); Jeremy Russell-Smith, "Studies in the Jungle: People, Fire and Monsoon Forest," in *Archaeological Research in Kakadu National Park*, ed. Rhys Jones (Canberra, Australia: Australian National Parks and Wildlife Service, 1985), 241–67; Jeremy Russell-Smith et al., "Aboriginal Resource Utilisation and Fire Management Practice in Western Arnhem Land, Monsoonal Northern Australia: Notes for Prehistory, Lessons for the Future," *Human Ecology* 25, no. 2 (1997): 159–95, http://dx.doi.org/10.1023/A:1021970021670.

52. Jeremy Russell-Smith et al., "Improving Estimates of Savanna Burning Emissions"; Jeremy Russell-Smith, Peter J. Whitehead, and Peter Cooke, *Culture, Ecology and Economy of Fire Management in Northern Australian Savannas: Rekindling the Wurrk Tradition* (Collingwood, Australia: CSIRO Publishing, 2009); Dean Yibarbuk et al., "Fire Ecology and Aboriginal Land Management in Central Arnhem Land, Northern Australia: A Tradition Of Ecosystem Management," *Journal of Biogeography* 28, no. 3 (2001): 325–43, http://dx.doi.org/10.1046/j.1365-2699.2001.00555.x.

53. Rosemary Hill, Adelaide Baird, and David Buchanan, "Aborigines and Fire in the Wet Tropics of Queensland, Australia: Ecosystem Management Across Cultures," *Society & Natural Resources* 12, no. 3 (1999): 205–23; Rosemary Hill et al., *Yalanji-Warranga Kaban: Yalanji People of the Rainforest Fire Management Book* (Queensland: Little Ramsay Press, 2004).

54. Latz and Griffin, "Changes in Aboriginal Land Management"; Peter Kenneth Latz and Jenny Green, *Bushfires & Bushtucker: Aboriginal Plant Use in Central Australia* (Alice Springs, Australia: IAD Press, 1995).

55. Tom Vigilante, "Analysis of Explorers' Records of Aboriginal Landscape Burning in the Kimberley Region of Western Australia," *Australian Geographical Studies* 39, no. 2 (2001): 135–55; Tom Vigilante, "The Ethnoecology of Landscape Burning Around Kalumburu Aboriginal Community, North Kimberley Region, Western Australia" (PhD diss., Charles Darwin University, 2004).

56. Sean Kerins, "The Future of Homelands/Outstations"; Kirsten Maclean, Cathy J. Robinson, and Oliver Costello, eds., *A National Framework to Report on the Benefits of Indigenous Cultural Fire Management* (Collingwood, Australia: CSIRO, 2018).

57. Hill, Baird, and Buchanan, "Aborigines and Fire in the Wet Tropics"; Tom Vigilante, "The Ethnoecology of Landscape"; Dean Yibarbuk et al., "Fire Ecology and Aboriginal Land Management."

58. Cherry Daniels, personal communication.

59. Enrique Salmón, "Kincentric Ecology."

60. Cook, Jackson, and Williams, "A Revolution in Northern Australian Fire Management"; Russell-Smith, Whitehead, and Cooke, *Culture, Ecology and Economy of Fire Management*.

61. Maclean, Robinson, and Costello, *A National Framework*.

62. Maclean, Robinson, and Costello, *A National Framework*; Michelle B. McKemey et al., "Cross-Cultural Monitoring of a Cultural Keystone Species Informs Revival of Indigenous Burning of Country in South-Eastern Australia," *Human Ecology* 47, no. 6 (2019): 893–904; Neale et al., "Walking Together."

63. Steffenson, *Fire Country*.

Chapter Three

African Sciences and Indigenous Knowledge Systems in the West African Ebola Crisis

TOLUWANI OLOKE AND OLUSEGUN SOETAN

Our study of the Ebola crisis in West Africa describes indigenous sciences and systems of knowledge inquiry within the purview of African health belief systems, and how these belief systems influence the complex relationships and stiff resistance of Africans to Euro-American health interventions in global health pandemics and epidemics. This study helps illustrate the claim that there are certain paradigms of health systems in the African belief systems that, despite the advent and development of Euro-American medical practices and medicines, have stood the test of time. We also argue that health communication campaigns that focus on the inefficacies of African scientific and medical methods communicate a fundamental disrespect of African epistemologies that ironically (1) obscures evidence and counterclaims supporting the efficacies of traditional African healing methods and scientific thoughts while (2) dissuading some Africans from accepting Euro-American treatments.

In line with Malcolm MacLachlan, who posits that every culture must be evolving, dynamic, and adaptive,[1] African health belief systems have also evolved over the years to overcome challenges. In the past, many arguments have been raised against the sustenance of traditional herbal medicine and in favor of orthodox Euro-American medicines, including that herbal medicines do not have recommended dosages and the concoctions are not refined or medically researched.[2] However, the years from the precolonial age to the present day have seen herbal medicine evolve to adequately compete with orthodox medicine. In modern times, herbal medicines are being repackaged the same way as orthodox medicines, to give them better presentations and added face value. For instance, herbal medicines are being assigned numbers from the Nigerian National Agency for Food and Drug Administration and Control (NAFDAC). The agency also ensures patient care and safety in relation to medication production and use by ensuring pharmacovigilance and compliance with all specifications. Not only does this grant herbal medicines acceptable legal status to Nigerian users and Nigerian medical practitioners in many African countries, but it also overrides the argument that they are not researched or refined.

African styles of inquiry into natural phenomena reflect a variegated practice that has helped multiple African societies to conceive the world in their own specific ways. Inquiry into the nature of matters follows different epistemological approaches in Africa. These epistemological approaches, on their own, are shaped by the worldviews and world senses of each African ethnic group and nation. In this chapter, we present two related case studies of how systems of medical inquiry and treatment differ between Euro-American and West African traditions, revealing the types of public arguments that can arise when these traditions come into conflict—as they did during the Ebola crisis of 2014–2015—along with some potential rhetorical solutions to that conflict.

When it comes to healing knowledge in Africa, there is evidence that supports the efficacy of herbal remedies and traditional scientific practices. Those who support indigenous African healing methods do so from informed perspectives, the majority being rural dwellers who rely on forest and fauna for their daily sustenance. Because of their agrarian lifestyles and remote locations, which are often far away from urban centers and towns, most rural dwellers evolved a healing system that tapped the curative potentials of leaves and herbs to survive. As a result of these hundreds—and in some cases thousands—of years of experience, they frequently uphold traditional healing methods as authentic and efficacious. On the other hand, city dwellers who have access to metropolitan infrastructures and modern knowledge uphold orthodox Euro-American medical knowledge as the most sustainable and reliable. The dichotomy between views and opinions about West African medical knowledge has, over the years, constituted a rhetorical debate about African epistemology.

Therefore, our case studies of Yoruba healing methods and therapeutic practices in southern Nigeria, as well as an Ebola-prevention campaign conducted in Guinea, Liberia, and Sierra Leone, stand as an invitation to engage more deeply and respectfully with African epistemological claims and scientific inquiries. While laboratory experimentation is the underpinning of Euro-American scientific thought, indigenous medical practices have been more integrated into a lifestyle than they have been isolated as a specialized field of knowledge. Healing knowledge is often treated as a secret and guided by the initiated and members of a practitioner's family. The lack of documentation and recording of African healing processes has emerged as a critical issue, and this lack of documentation was driven in the first place by a fundamental lack of respect for African scientific traditions: thus, the central irony of the rhetorical situations we examine here in West Africa.

From these case studies, two notable insights emerge with respect to our research focus: First, the agelong cultural beliefs and practices of a people cannot be separated from the people because the beliefs and practices form their living and existence. African health systems and beliefs have a survived great onslaught of Euro-American propaganda and the accompanying challenges that arise when Euro-American science and medicine push up against indigenous/local knowledge systems. Second, it is easier and less complicated to negotiate this conflict when the paradigms of the indigenous culture are understood, respected, and gradually implement a change in the long run. Such paradigms as identified include religion, language, and social constructs. These insights will be further developed in subsequent sections of this study.

West African Rhetorics of Science

Science is associated with European modernity and other civilizations from the East; however, historians of science are coming to recognize that every society produces its own science.[3] Stylistically, scientific inquiry varies from one population to another, and they follow different routes cross-culturally. In its minimalist form, science is a cyclical process of resolving human problems. These problems could be social, material, biological, or physical. When people seek means of solving their own crisis, they are performing scientific exploration. According to Kuhn, "Scientific communities conform to certain norms until a crisis challenges them, forcing the emergence of a new paradigm that resolves the crisis."[4] The understanding here is not a universal assumption about science; instead, it is a multicultural proposition that undergirds the multiple ways of doing science. If science is primarily concerned with "certifiable knowledge,"[5] that is, statements of regularity that are empirically confirmable and logically consistent, then precolonial African societies are scientific communities.

Continental Africa was and is, however, still regarded as a consumer rather than a producer of evidence-based science. Nevertheless, Africans have developed and sustained their indigenous technologies and perfected autonomous scientific approaches that serve their needs. Traditional African sciences evolved on their own and cultivated peculiar inquiry protocols different from those prototypes developed by Euro-American scientists. The difference in what constitutes science to Africans and the rest of the world, especially the Global North, is not about a significant difference in the cognitive abilities of the people from the two hemispheres (Global North and the Global South); instead, it is about positionality and various cultural formations. No matter their global rating, no culture is unscientific: All cultures and places have their own science and modes of scientific inquiry. However, the global socioeconomic system praises knowledge created by economically developed countries while dismissing epistemological pieces of evidence from Global South countries such as those in Africa.[6]

The failure to recognize these accomplishments has in large part to do with the way African sciences are talked and written about—in other words, their rhetoric. One rhetorical problem lies in definitions of science literacy, which colonizers believed Africans lacked. Western societies documented their ideas and circulated them beyond their boundaries. By doing so, they grew their knowledge and preserved their patent rights. The ability to document ideas in written forms helped Westerners to lay claims to knowledge creation because Euro-American science is based on an academic and literate transmission of that knowledge, including ideas appropriated from other places and cultures; meanwhile, traditional African knowledge is dominantly passed on orally from one generation to the next by elders.[7] On the African continent, almost all societies used oral traditions to document ideas and communicate processes from one generation to the other.[8] Pieces of analyzed archeological ruins excavated from sites in African countries have shown that African communities and societies were scientific in their own ways. For example, chemical analysis of glass beads from Igbo Olokun, Ile-Ife (southwestern Nigeria), has shed new light on glass production technology among the Yoruba.[9] A related recent study also revealed that many nations still communicate records of their ancestors' environmental information, observations, and applications such as eclipses, sea level changes, and other scientific observations as shared knowledge passed down from one generation to the next until the present day.[10]

A second error of judgment regarding the rhetoric of scientific knowledge in Africa comes from the fact that many African communities do not separate science as a specialized body of knowledge. The fundamentals of scientific inquiry in the Euro-American tradition were categorized as subjects and developed for study. For instance, chemistry, the science of matter and its transformations from one form to another,[11] was codified as a discipline nearly a thousand years ago and is still studied in an orthodox format across all levels of formal education, a system that projects the discipline as advanced and methodical. By contrast, from topics relating to hygiene, socialization, and education to trade and health, scientific inquiry is embedded in everyday African technical and cultural practices and philosophies.[12] These inquiries not only involve casual observations of the immediate environment, but also a thorough reading of situations and, possibly, empirical study of phenomena. For example, the Nigerian grandfather of one of this chapter's authors, Soetan, was a healer and a farmer. He would experiment with curative leaves and herbal mixtures to find antidotes for diseases and illnesses, but he was never referred to as a chemist or a pharmacist. His experimental endeavors in finding antidotes and herbal medicines for illnesses and diseases was certainly a form of traditional science but was not acknowledged as such by orthodox scientific authorities in the region. On this point, we must observe that Euro-American conceptions of "scientist" or "scientific" necessarily had no meaning for precolonial Africans. Knowledge creation was a communal enterprise as people came together to create mechanisms for solving their problems, and in the end, they gave credit to everybody. This, perhaps, explains why many African stories, proverbs, and myths do not have known authors: they are regarded as the collective wisdom of the community.

To think through the production of scientific knowledge in West Africa, therefore, we must be ready to go deeply into the culture to extrapolate the science and scientific facts buried in them. We cannot understand West Africans' perception of science if we do not holistically scrutinize their communal cultural practices and processes. In that regard, African languages are critical to understanding what science means to Africans. According to Mavhunga, "Everyday language expresses realities and imaginations at the intersection of African inventions and inbound idioms and thus testifies to the creativities of Africans who strategically deploy them."[13] African languages are the archive that stores and passes on scientific ideas from one generation to another. In the multiplicity of African proverbs and idiomatic expressions, empirical statements and technological ideas are coded for didactic purposes, even though they were never considered scientific knowledge. For example, there is a Yoruba idiom that says, "Lálá tó ròkè, ilẹ̀ ló ń bọ̀" (any matter/object that goes up must come down), a cultural principle that is clearly built on the law of gravity. However, the Yoruba have not codified gravitational rules and principles in a written statement, as in Newton's Law of Universal Gravitation. What this example suggests is that bits of African scientific knowledge are hidden in general cultural forms and everyday life processes; and, rhetorically speaking, African scientific methods have relied on a broader cultural mode of delivery than Euro-American science. A wider view of scientific practice is thus required to comprehend them.

West African Health Science: The Yoruba of Nigeria

The Yoruba of Nigeria can be conceived of as a language, as a religion, or as a nation (people). For this study, we will adopt the perception of the Yoruba as a people/nation. The Yoruba people of southwest Nigeria live predominantly in seven states in Nigeria—Ogun, Oyo, Lagos, Ekiti, Osun, Ondo, and Kwara—and they rank among the most educated in the country. Historically, the Yoruba nation controlled a vast empire that had its headquarters in Oyo and that extended to the present-day republics of Benin and Togo.[14]

Like many African societies, the Yoruba people have evolved their own scientific methods. The Yoruba style of inquiring into natural phenomena relies on visual observation and other sensory perceptions. However, these methods are firmly rooted in the group's philosophy and cosmogony. Therefore, a deeply embedded and broadly circumspect approach is required to fully appreciate Yoruba health science practices. The claims made in this case study are accordingly based on one of our (Soetan's) life experiences in and study of Yoruba cultural practices. While this case study approach follows current recommendations by Agboka and others for decolonial ethnographic work in non-Euro-American technoscientific contexts,[15] it also goes significantly beyond them, as many of the practices described here could not possibly be grasped or explained by an outsider who spent a few months or even years among the Yoruba: on the contrary, the level of understanding demonstrated in the following narrative case study requires cultural imbrication as well as critical reflection.

The Yoruba conceive our world as a space where matter regularly interact with one another, and that all natural elements—trees, rivers, birds, animals, rocks, and celestial matters—have two names: a common name and a hidden cognomen. This belief system is the basis for understanding Yoruba science and technology, including health science. Instances of scientific innovation in traditional Africa are most observable in medical and health care practices. To understand African epistemology as it relates to science and technology, one must begin with historical health practices that sustained the people in precolonial times and are part and parcel of African postcolonial experiences.

Adeoye has explained that the Yoruba people from ancient times understood the inevitability of death.[16] Nevertheless, they also believed that certain incidences and mishaps can lead to untimely death and, as such, it became important to understand how to preserve health and life using herbs and metaphysical powers.[17] As mentioned in the introduction to this chapter, there are three paradigms within which these healing systems can be understood, paradigms inherited from the precolonial age. The first paradigm is religion and the belief in metaphysical powers and rituals in healing processes. In fact, Aderibigbe goes so far as to argue that "the belief in medicine and magic constitutes the fourth segment of the belief structure in African traditional religions."[18] The severity of an ailment such as Ebola would determine if the traditional healers (physicians) would use only herbal mixtures or involve metaphysical powers through divinations and rituals. Diagnosis is done through divinations and monitoring of symptoms presented by the patient. These practices demonstrate that Yoruba traditional herbalists examine their professions through the lens of religion. It is possible that the people as a nation understand their states of health and wellness through the lens of religion as well. As Angellar Manguvo and Benford Mafuvadze concluded from their study of the Ebola epidemic in Yoruba territory and neighboring areas, "Overall, reports of attempts to cure Ebola through traditional and spiritual means highlight the need for increased cooperation between traditional healers, spiritual healers, and trained health personnel."[19]

Second, the concept of wellness is construed through the paradigm of social well-being. Wellness does not wholly indicate physical and mental health but focuses more on social well-being and a state of being protected from epidemics such as Ebola by charms and higher forces that are constantly appeased through rituals. Wellness could be understood within the Yoruba people as other social, self-actualizing needs such as success, good harvest, protection, wealth, and others. In other words, the health and well-being of people is not handled with levity, but it requires very careful study and observation over many years to become well-versed in handling health-related cases.

The third paradigm is linguistic and socio-mythical. This is reflected a little in the way that the English word *wellness* is interpreted as different from being *healthy*. In a deeper sense, however, the categorization of different illnesses influences how ailments are referred to among the Yoruba people. Severe ailments such as Ebola are called *aisan* (illness) while less complicated ailments are referred to as *arun* (disease). While *arun* could be treated with herbal treatments, *aisan* requires spiritual/metaphysical interventions because they are not considered physically treatable.[20] In explaining the socio-mythical approach, Ọládélé Caleb Orímóògùnjẹ́ has argued that "some of the Yorùbá verbal arts are mythical allusions in which myth can be used as a tool to unveil the hidden issues. Therefore, it paves the way for getting acquainted with valid information on issues like how diseases are caused, prevented and treated in the context of cultural tradition."[21] The linguistic and socio-mythical paradigm highlights the link between verbal culture and the Yoruba traditional health system. For example, in the height of the COVID-19 pandemic, most public education in some Yoruba-speaking states appropriated indigenous knowledge of infectious disease management to educate the people. The cultural appropriation made use of terms such as *akọ ìgbóná* (high fever), which is generally associated with *ṣànpọ́nná* (smallpox) to teach the masses about the symptoms of the COVID-19 virus.

Yoruba herbal medicine is therefore both a cultural practice and a science. First, as a cultural practice, it draws from the universal knowledge (the undercurrent of Yoruba philosophy) that all naturally occurring elements are composed of either positive or negative energy, and that when these energies are combined they form a reaction. The concept of positive versus negative energy is expressed as a duality—*akọ/abo*—(male⁺/female⁻). Second, as a science, Yoruba herbal medicine requires that somebody be knowledgeable enough to distinguish between positively charged plants and those that carry negative charges. The Yoruba indigenous pharmacopeia means that the healer or the traditional medical doctor is both a cultural mediator and a scientist at the same time. The undercurrent of their trade follows the cultural logic of the people regarding holistic therapy. The healer is considered a professional if they can "wa egbò dẹ́kun" (apply a plant root to end a disease), which implies is that the healer treats causes of ailments and diseases and not just the symptoms.

We can understand African medical science's specificity by paying attention to the so-called "illiterate" and "unorthodox" medical doctors, people with knowledge of herbal plants and curative formulas. For example, among the Yoruba of Western Nigeria, the *onísẹ̀gùn*, not the *babalawo* (oracular herbalist and charm-maker), is responsible for curing diseases and ailments, and they rely on the use of herbs, medicinal plants, and the exoteric power of incantations. There is also the traditional bonesetter, who handles orthopedic ailments. We will consider each of these two healing practices in turn, observing the interrelation of religious, social, and socio-linguistic/mythical factors in each.

One of the first "professions" recognized by precolonial Yoruba people was health care, which is otherwise known as herbal medicine, the main practitioners of which were known as *onísẹ̀gùn*.[22] By combining medicinal properties from leaves and plants, the *onísẹ̀gùn* heals sick people and cures diseases. Despite the charges against African traditional medical practices as unsafe, many Yoruba continue to patronize traditional healers while preferring alternative herbal medicines to modern hospitals and clinics. West Africans continue to choose traditional healers over modern hospitals because traditional medicine is "the oldest form of health care system that has stood the test of time for Africans and it is an ancient and culture-bound method of healing through which traditional healers attempt to reconnect the social and emotional equilibrium of patients based on community rules and relationships unlike medical doctors that only treat diseases in patients."[23]

The Yoruba idea of holistic healing is enshrined in the proverb "pátápáta là ń fójú, kùnàkuna là ń dẹ́tẹ̀, ojú à-fọ́ọ̀fọ́-tán, ìjà níí dá sílẹ̀" (total blindness is better than a partial sight; if one wants to be leprous, it better be complete; partial sight brews acrimony). Yoruba herbal medicine is designed to heal all diseases and cure all ailments, returning the body to a state of health, and that is why its practice is different from those of other African societies, and from Euro-American medicine. Besides combining leaves to make a concoction for drinking, sometimes doctors make incisions on the bodies of their patients to infuse chemical compounds in their blood or as a form of inoculation against infectious diseases. While these healing methods do not follow orthodox medical practices, based on inherited knowledge and the experiences of the authors, they have been proven to be useful.

A major preoccupation of the *oníṣègùn*, which demonstrates the interweaving of religious, social, and mythical/linguistic paradigms, is *Ìgbóná* or *olóde* (smallpox). Smallpox is an acute contagious disease caused by the variola virus according to Western medical literature. However, among the Yoruba people of Nigeria, smallpox is considered an epidemic caused by a deity Ọbalúayé or ṣànpọ̀nná. Thus, the etiology and treatment plans differ considerably from the western approach to curing smallpox. Instead of vaccination, the Yoruba *oníṣègùn* uses herbs and concoctions to treat the patient. For the most part, the patient is kept away from other members of the family and community. Usually, the treatment plan is twofold: a sacrifice and an herbal remedy. The sacrifice is an offering of atonement to Ọbalúayé/ṣànpọ̀nná, the deity of smallpox, to abate his wrath. The herbal treatment varies from one *oníṣègùn* to another. However, common herbs and leaves for making the smallpox antidote are *tàgíìrì*, *orí*, *ọṣe-dúdú*, and *ejìnrìn* (Christmas melon, shea butter, black soap, and bitter gourd). The *oníṣègùn* mixes the medicinal materials together to make a paste that they apply to the rashes. Also, the Christmas melon could be soaked in rainwater to make a concoction for the patient to drink. In most cases, the smallpox rashes heal in as little as three days depending on the consistency of treatment.

Besides herbal medicine, there are other ways of seeking protection against diseases, misfortune, and danger. To seek protection from evil and dangers, the Yoruba people make charms and amulets, which "refer to a belief system or a natural science that bundles synergic exchange between the transcendental heavens and the immanent earth hidden in scientific forms held secret among the initiates schooled in the communicative codes."[24] These communicative codes are otherwise referred to as incantations, as mentioned earlier in the discussion of the religious and sociolinguistic dimensions of Yoruba medicine. To a Euro-American scientist, charm-making is not a scientific endeavor; instead, it is viewed as a cultural practice that lacks any empirical process. However, apart from it being fundamental to the ways the Yoruba make sense of the world and create the world for themselves, it is essentially an epistemological undertaking that appropriates multiple knowledge sources. Charms and amulets work through the appropriation of chemical matters in plants and bundling these chemical properties with other animal energy sources. This is because the world, according to the Yoruba people, is populated by human beings, animals, plants, and other nonliving objects that continuously interact with one another.

At the heart of charm-making is serious research into the chemical properties of matters and their ability to combine. Let us take, for example, the bulletproof charm (*ayẹta*), which is designed to repel the projectiles from the local Dane guns, and which warriors relied on during wars. Unlike the Euro-American bulletproof vests manufactured in big companies and worn by security agencies, the *ayẹta* is not a vest, but an amulet that is worn underneath clothes. This charm is usually made from plants and animal parts that are crushed together, or burned into powder, and then transfused into the blood through an incision. Once the powder gets into the blood, it makes the muscles of the user repel bullets. Whenever a bullet is shot at the user, their muscles contract together to form a protective barrier that repels the bullet.

To further contextualize African inquiry into natural phenomena, let us investigate the chemical composition of an *ayẹta* charm. Doing so will provide us more critical insights into the nature of African science, which is methodically different from Euro-American thinking. There are more than one hundred formulas for making the *ayẹta* charm, but let us consider one that has the four leaves listed below as its primary reactants:

1. *Ewé àlùpàídà* (*Uraria picta*)

2. *Ewé ogbe àkùkọ* (*Heliotropium boraginaceae*)

3. *Atare àjà* (*Aframomum melegueta zingiberaceae*)

4. *Ewé Àbámodá* (*Bryophyllum pinnatum*)

To the untrained eye, the four reactants in the formulae above are just typical leaves that carry no specific properties to repel bullets. However, to the charm-maker, these reactants have properties that, when combined, can deflect bullets. In Yoruba botany, *Uraria picta* causes transformation that, if combined with other leaves, can make objects disappear. Here, too, it is combined with *Bryophyllum pinnatum*, a plant that has hypnotizing properties, and *Aframomum melegueta zingiberaceae* acts as the catalyst to speed up the reaction. These four reactants will combine with other plants and/or animal parts to make the charm.

Furthermore, once the charm is finalized, it is tested for reliability. To test the efficacy of the charm, the amulet is usually worn on the neck of a goat or dog. Once the amulet is strapped to the tethered animal, the user goes on to shoot at the animal. If the bullet hits the animal and it dies, it means the charm has failed. However, if the bullet hits the animal and it suffers no wound, then the charm is valid.

If we consider the process that went into charm-making, we realize that it follows a method. Although this method may differ significantly from Euro-American scientific methods, it is still a system that conducts a reliability test. In modern Euro-American practice, scientists still test many medicines on animals before they embark on human testing and mass production.[25] The test is designed to ensure that the new products are safe for public use and are reliable. Similarly, in the local context of charm-making, the charm-makers use animals to test their amulets to confirm that they are safe for the user to use. What these similarities point our attention to are the various ways people across cultures practice science and what constitutes scientific inquiries to different populations and geographies. If one pays a significant amount of attention to the technology of charm-making among the Yoruba people, one will realize that the knowledge of botany is critical to the endeavor; not only that, but there is also an enormous understanding of matter and their abilities to combine in the science of charm-making.

With the growing demand for traditional medicine in African countries,[26] many Yoruba people are turning to traditional bonesetters. People often claim that attending orthodox orthopedic clinics is not always beneficial because it sometimes results in amputation and death. As a result, they trust the traditional bonesetters and patronize them. In practice, bonesetting in African parlance is different from Euro-American orthodox orthopedic practices. Without the sophistication of X-ray machines and other diagnostic equipment, bonesetters in many African communities have successfully healed broken bones and corrected complicated fractures. In Nigeria, and among the Yoruba people, bonesetting is the dedicated practice of specific individuals. Some families are renowned for their expertise at setting bones and carrying out other orthopedic surgeries using herbs and animal parts. Bonesetting, many times, uses a reverse engineering method to cure broken bones. In treating fractured bones, mostly simple fractures, the bonesetter begins their healing process using a fowl as a specimen. They break the leg of a fowl at the exact location as that of the human leg that they intend to cure. They support the fractured leg with wood splints and bamboo sticks, and commence treatment. Every day, a specific dose of herbal concoction is used to give the bird's leg a sponge bath. As the bonesetter treats the bird, they apply the same herbal remedy to the human leg. Occasionally, the bonesetter chants incantations and offer sacrifices to the deities. After a while, the fowl leg heals, and the bird regains the use of its leg once again. At the same time, the patient's leg heals, and they are discharged to go home.

This traditional African orthopedic treatment, no doubt, contradicts the conventional Euro-American orthopedic knowledge that relies on the use of plaster of paris, metal support, and amputation, among other methods, to treat fractured bones. The method and the concoction that the traditional bonesetter uses can be dismissed as unscientific and primitive to the Euro-American scientist. Nevertheless, indigenous African science follows its scientific pathways and, if well-scrutinized, includes global medical ideas explained in vernacular cultures. A curious observer of the Yoruba bonesetter would realize that the practitioner always includes two main medicinal plants in their concoction: *ewé atò* (*Chasmanthera dependens*) and *òrí* (shea butter, made from *Vitellaria paradoxa*). These herbal plants contain medicinal properties that aid bone union. Extracts of *Chasmanthera dependens* have analgesic potential, and also serve as an anti-inflammatory agent. By applying a solution of *Chasmanthera dependens* on a fractured leg, the Yoruba bonesetter aims to eliminate edema, thus alleviating pain. Also, the shea butter, which is an anti-inflammatory balm, helps to soften the skin so that the herbal concoction can permeate the cutaneous/subcutaneous layers of the leg for effective treatment.[27]

What we have seen from the case study of Yoruba herbal medicine, charm-making, and bonesetting practices is that the agelong cultural beliefs and practices of a people are not just a set of rules that they live by, but constitute their living and existence. As Toyin Adefolaju has said, "That a structure exists presupposes its continued functioning and therefore, relevance to the existence and survival of the whole system. Traditional medicine has been developed to enable people to meet their health needs. This health system has survived great onslaught from Western propaganda and the modernization process. However the people still patronize it, in spite of the diminishing and derogative status accorded it. This suggests its functionality and continued relevance to the health needs of the people."[28]

There has also been a recent surge in the use of combination therapies using conventional orthodox medicine and traditional medicine. This hybridization has come about partly because traditional medicine treatments are cheaper, have religious and cultural undertones,[29] and are perceived to be more trustworthy because they are agelong practices that were effective before the advent of conventional/orthodox medicine. It is further important to note that traditional healing practices are being embraced by medical practitioners who have a Western education as well as educated traditionalists in societies such as the Yoruba. For example, the father of Oloke, one of the authors, who is a college graduate with two master's degrees, would always administer herbal concoctions made from an herbal leaf known as *ejirin* (*Momordica charantia*) as a prophylaxis treatment for malaria every month. The few times that the family visited a hospital for treatment were times when they had complications from typhoid fever. In such instances, they needed to be treated for typhoid at the hospitals.

We can conclude based on the intense cultural integration and hybridization observed in our case study of Yoruba health-science practices that health communication strategies in this community—and others like it, as traditional healing practices are shared widely across this region of Africa[30]—must also engage a situated and hybrid approach to be effective. That conclusion is borne out by our second case study on the Ebola crisis in West Africa.

Euro-American and African Indigenous Science and Medicine: The Ebola Outbreak

As mentioned above, traditional herbal medicines are rapidly gaining ground and recognition as a complement to orthodox medicine. Even though Euro-American medicine is now widely accepted as the norm in many African countries, studies have shown that over 80 percent of Africans, their varying levels of academic achievements notwithstanding, still visit traditional healers for health consultations.[31] This is mainly because orthodox medical treatment costs are exorbitant and there are also certain medical conditions with no medical cure within orthodox medicine that traditional medicine has successfully provided cures for.[32] An example of such a medical condition is the treatment of piles, which requires surgery to totally remove the hemorrhoids (hemorrhoidectomy); meanwhile, there are herbal concoctions that adequately treat these conditions without surgery.[33]

A major conflict that is yet to be resolved, however, is that the medical establishment in West Africa—at least publicly—has shown little to no respect for agelong traditional science/healing methods.[34] This position was forced to shift during the most recent Ebola virus disease outbreak in West Africa in 2014–2015, which recorded the highest number of cases and deaths compared to previous outbreaks of Ebola virus disease.

One of us, Oloke, had the privilege of working closely with orthodox health care practitioners for eight months, and also organized health awareness campaigns as part of the African Union Support for the Ebola Outbreak in West Africa (ASEOWA) in Liberia, Sierra Leone, and Guinea under the auspices of the African Union. The observations reported here draw from a corpus of 17 video interviews, 114 situation reports (mostly daily reports of communication interventions and efforts submitted to the headquarter office), and pictures and video clips from the communication awareness and education programs (see figure 3.1). The interviews were mostly unstructured and open-ended, but some questions focused on why locals were refusing to report new cases of Ebola or suspected cases to the medical teams. This question was pivotal in the sense that it determined the course of other questions based on the responses of interviewers. The responses mostly reflected the preference for traditional African medicine and cures through herbalists because that has been the long-standing, most trusted health care system that they have known for many generations.

Even though most ASEOWA campaign messages strongly encouraged everyone to seek medical care in hospitals, doctors and other health care practitioners realized that they needed to work in collaboration with traditional herbalists to curb the spread of Ebola. Several traditionalists interviewed claimed to have the cure to Ebola, and their claims were strongly supported by the testimonials of a significant number of patients. While there were no statistical data to show the number of patients that were allegedly cured of Ebola by the traditionalists, it is important to note that the rural areas where they worked had little to no access to orthodox medical care and were at the same time the epicenters of the Ebola outbreak. Such rural areas used traditional healers' homes as their "emergency rooms" in cases of critically urgent health situations.

Figure 3.1. Participants at the Ebola awareness training organized by ASEOWA in Grand Cape Mount, Liberia. *Source*: Photo by Toluwani Oloke.

In Liberia, for example, areas such as Grand Cape Mount County had very limited access to health care, which meant that suspected Ebola cases were grossly underreported to the emergency coordinating centers because people resorted to their traditional healers for care. Field visits and situation report statements revealed various claims and testimonials of Ebola treatment and cure by traditional healers in these rural regions. According to an interview with a local resident whose family members contracted the Ebola virus, his brother, who worked as a traditional healer, successfully managed patients with early symptoms of the Ebola virus using the same herbal concoctions that were used to treat other types of fever. Considering that Ebola virus symptoms at the first phase of infection are the same as the symptoms for other illnesses like malaria, such as headache, fever, and extreme tiredness, this claim is logical. Also, the survival rate from the Ebola virus disease, if reported early and treated before the hemorrhagic stage sets in, is higher.

In personal interactions with the members of the international health agency volunteers during the Ebola outbreak in Liberia and Sierra Leone, the major criticisms against the approach of the traditional healers were poor IPC (infection, prevention, and control) protocols and poor knowledge about the dosage of their concoctions. These criticisms are valid, but they did not strongly falsify the claims of Ebola cure by both the traditional healers and their patients. In Sierra Leone, communities such as Port Loko, Kenema, Koinadugu, and Bombali recorded high cases of the Ebola virus disease outbreak, and in most of these cases, traditional healers were consulted by a number of survivors who claimed during personal interviews to been cured. There were other claims that only cases that resulted in complications—for example, bleeding through orifices or breathing problems—got transferred as emergency cases to the orthodox treatment units. A community head in Sierra Leone insisted that these cases were referred because of limited basic infrastructure and facilities in rural areas that were readily available in urban areas where the Ebola Treatment Units (ETUs) were located.

The threat of "weaponized" Ebola in the public mind further complicated the stiff response and hostility towards orthodox medical intervention in the three countries that were the epicenter of the Ebola outbreak. Ebola was widely rumored among the locals to be an artificial virus created by the Western communities and injected into "bush meats," an African delicacy, as a way to eradicate the entire African race. As preposterous as this rumor sounds, it was widely spread and believed by many of the locals the ASEOWA team interviewed, which explained their initial hostile reactions to the interventions of the international responders. A much stronger rumor peddled by some survivors of the Ebola outbreak was the claim that they survived Ebola because they refused to take the "yellow pills" administered to them in the treatment units by the medical doctors. According to the unfounded rumors, the yellow pill was meant to end the lives of patients who took the pills. Apparently, the pill was one of the vitamin supplements that were administered to patients in the ETUs to help their immunity and improve their chances of survival. This situation further strengthened the resolute stance of the locals to seek a cure with their traditional healers instead of the international medical responders. One of us, Oloke, experienced a mob attack by the locals in Guinea, who strongly resisted the presence of the ASEOWA medical team in their midst. A similar attack occurred in West Point, Monrovia, Liberia, during a sensitization campaign. The attacks were ways to express displeasure at the presence of medical interventions in which the people had a lot of initial mistrust.

The ASEOWA Ebola communication response/intervention started out purely with the orthodox medical perspective: for example, it aimed to convince the local people to seek medical interventions at ETUs, posted only medical personnel who were Africans to the treatment units to build trust, and engaged with local health care centers to improve quality of care. However, this approach did not yield the expected result of reduction in infection and an increased case reporting rate. So the communication team reviewed and strategized, and decided to meet with the local people at town hall meetings (see figures 3.2 and 3.3). The meetings revealed that people would rather work with traditional herbalists because, according to them, their symptoms improved after consulting with the herbalists, without orthodox intervention. Even though these claims were not accepted by the communication teams, it pointed to the fact that a hybrid approach would probably work best. From a strategic communication perspective, the communication team then partnered with indigenous celebrities, the police force, and selected local nongovernmental organizations (NGOs) to convince the people to accept orthodox medical interventions as complementary to traditional treatments (see figure 3.4).

What we conclude from this case study is that traditional medicine proved effective during the Ebola outbreak in Liberia, Sierra Leone, and Guinea for patients who self-reported early to traditional healers. The case

Figure 3.2. Town Hall Meeting 1. *Source*: Photo by Toluwani Oloke.

Figure 3.3. Town Hall Meeting 2. *Source*: Photo by Toluwani Oloke.

study also calls for a hybrid rhetorical approach to health care and pandemic management in West African traditional societies. Our experiences confirmed Claire Munoz Parry's study of how the people's science helped

Figure 3.4. African Union ASEOWA joint awareness collaboration with Liberian Police Force. *Source*: Photo by Toluwani Oloke.

end the Ebola epidemic, which concluded that the Ebola outbreak was brought to a halt not by the intervention of international responders but by local health practitioners and the people themselves.[35]

Conclusion

Since colonization, health care systems in the Global South, especially in Africa, have multiplied, not reduced.[36] Neither traditional nor orthodox medicine is adequately catering to the health care needs of the African people. A synergy of approaches appears to be better. Ironically, in spite of movements geared towards demystifying traditional herbal medicine and its claims of metaphysical interventions, contemporary traditional herbal treatments are surging. Traditional healing systems have been professionalized in recent times and there are new legally recognized professional organizations formed by Yoruba traditional healing practitioners. For a culture to survive, it must be dynamic and should adapt to the evolving environment.[37] The Yoruba culture of traditional healing is also evolving, with herbal medicines being packaged and administered in recommended dosages.

To sum up the insights from the Yoruba and Ebola case studies for global health communication: The consideration of traditional and religious practices remains critical to our understanding of transmission dynamics and subsequent control of highly infectious diseases.[38] Also, health communication strategies that don't respect these practices ironically produce resistance to Euro-American alternatives, while simultaneously cutting off chances to gather important evidence on the effectiveness of traditional treatments. A hybrid rhetorical strategy can reopen lines of communication for the benefit of everyone involved.

So many important questions about West African health sciences remain: How, for instance, did Yoruba society resolve the pandemics of the 18th and the 19th centuries? What were the maternal health practices among the Akan people in present-day Ghana before the European colonization of their territory? How did West African communities, especially Nigeria, manage the most recent Ebola outbreak of 2014? African scientific communities have operationalized science differently from one region to another, based on their particular needs. But we can only adequately understand and integrate these sciences once we have stopped filtering the answers to our research questions through the dominant Euro-American paradigm. We can take a step in that direction by pursuing deeper, broader, more rhetorical and collaborative engagements with African scientific practitioners.

Notes

1. Malcolm MacLachlan, *Culture and Health: A Critical Perspective towards Global Health* (New York: John Wiley & Sons, 2006).

2. Ali Arazeem Abdullahi, "Trends and Challenges of Traditional Medicine in Africa," *African Journal of Traditional, Complementary, and Alternative Medicines* 8, no. 5 (2011): 115–16, 10.4314/ajtcam.v8i5S.5.

3. Oseni Taiwo Afisi, "Is African Science True Science? Reflections on the Methods of African Science," *Filosofia Theoretica* 5, no. 1 (2016), https://doi.org/10.4314/ft.v5i1.5.

4. Thomas S. Kuhn, *"The Structure of Scientific Revolutions"* (Chicago: University of Chicago Press, 1962).

5. Robert K. Merton, *The Sociology of Science: Theoretical and Empirical Investigations* (Chicago: University of Chicago Press, 1973).

6. Merton, *The Sociology of Science*.

7. Fulvio Mazzocchi, "Western Science and Traditional Knowledge: Despite Their Variations, Different Forms of Knowledge Can Learn from Each Other," *EMBO Reports* 7, no. 5 (2006): 463–66., https://doi.org/10.1038/sj.embor.7400693.

8. "Historical Archaeology in Africa: Representation, Social Memory, and Oral Traditions," *Choice Reviews Online* 44, no. 12 (2007).

9. Abidemi Babatunde Babalola et al., "Chemical Analysis of Glass Beads from Igbo Olokun, Ile-Ife (SW Nigeria): New Light on Raw Materials, Production, and Interregional Interactions," *Journal of Archaeological Science* 90 (2018): 92–105, https://doi.org/10.1016/j.jas.2017.12.005.

10. Elizabeth Rasekoala and Lindy Orthia, "Anti-Racist Science Communication Starts with Recognising Its Globally Diverse Historical Footprint," *Impact of Social Sciences* (blog), July 1, 2020, https://blogs.lse.ac.uk/impactofsocialsciences/2020/07/01/anti-racist-science-communication-starts-with-recognising-its-globally-diverse-historical-footprint.

11. John W. Moore and Conrad L. Stanitski, *Chemistry: The Molecular Science* (Boston: Cengage Learning, 2014).

12. Hassan O. Kaya and Yonah N. Seleti, "African Indigenous Knowledge Systems and Relevance of Higher Education in South Africa," *International Education Journal: Comparative Perspectives* 12, no. 1 (2013): 36–38.

13. Clapperton Chakanetsa Mavhunga, *The Mobile Workshop: The Tsetse Fly and African Knowledge Production* (Cambridge, MA: MIT Press, 2018).

14. Jeremy Seymour Eades, *The Yoruba Today* (Cambridge: Cambridge University Press, 1980).

15. Godwin Y. Agboka, "Decolonial Methodologies: Social Justice Perspectives in Intercultural Technical Communication Research," *Journal of Technical Writing and Communication* 44, no. 3 (2014): 297–327; Jennifer Manning, "A Decolonial Feminist Ethnography: Empowerment, Ethics and Epistemology," in *Empowering Methodologies in Organisational and Social Research*, 39–54 (India: Routledge, 2022); Juno Salazar Parreñas, "From Decolonial Indigenous Knowledges to Vernacular Ideas in Southeast Asia," *History and Theory* 59, no. 3 (2020): 413–20.

16. C. L. Adeoye, *Asa Ati Ise Yoruba* (Ibadan, Nigeria: Oxford University Press, 1979), 119.

17. Adeoye, *Asa Ati Ise Yoruba*, 119.

18. I. S. Aderibigbe, "The Traditional Healing System among the Yoruba," in *Traditional and Modern Health Systems in Nigeria*, ed. Toylin Falola and Matthew H. Heaton (Trenton, NJ: Africa World Press, 2006), 365–80.

19. Angellar Manguvo and Benford Mafuvadze, "The Impact of Traditional and Religious Practices on the Spread of Ebola in West Africa: Time for a Strategic Shift," *The Pan African Medical Journal* 22 (2015), https://doi.org/10.11694/pamj.supp.2015.22.1.6190.

20. Aderibigbe, "The Traditional Healing System among the Yoruba."

21. Ọládélé Caleb Orímóògùnjẹ́, "The Yorùbá Indigenous Psychotherapeutic Healing System: A Case Study of Oríkì," *International Journal of Humanities and Cultural Studies (IJHCS)* 2, no. 4 (2016): 856–65.

22. Orímóògùnjẹ́, "The Yorùbá Indigenous Psychotherapeutic Healing System," 856–65.

23. Abdullahi, "Trends and Challenges," 115.

24. Olusegun Soetan, "Charms and Amulets," in *Culture and Customs of the Yoruba*, ed. Toyin Falola and Akintunde Akinyemi (Austin: Pan-African University Press, 2017), 205–13.

25. Hugh LaFollette and Niall Shanks, *Brute Science: Dilemmas of Animal Experimentation* (London: Routledge, 2020).

26. L. Carpentier et al., "Choice of Traditional or Modern Treatment in West Burkina Faso," *World Health Forum* 16, no. 2 (1995): 198–202.

27. Nandini Verma et al., "Anti-Inflammatory Effects of Shea Butter through Inhibition of INOS, COX-2, and Cytokines via the Nf-Kb Pathway in Lps-Activated J774 Macrophage Cells," *Journal of Complementary and Integrative Medicine* 9, no. 1 (2012): 1–11, https://doi.org/10.1515/1553-3840.1574; Oumar Thioune et al., "Contribution of Nanotechnology In the Improvement of the Anti-Inflammatory Activity of Shea Butter.," *American Journal of PharmTech Research* 9, no. 6 (2019): 242–53, https://doi.org/10.46624/ajptr.2019.v9.i6.021.

28. Toyin Adefolaju, "The Dynamics and Changing Structure of Traditional Healing System in Nigeria," *International Journal of Health Research* 4, no. 2 (2011): 100.

29. Peter Bai James et al., "Pattern of Health Care Utilization and Traditional and Complementary Medicine Use among Ebola Survivors in Sierra Leone," *PLoS ONE* 14, no. 9 (2019), https://doi.org/10.1371/journal.pone.0223068.

30. Abdullahi, "Trends and Challenges."

31. Peter Bai James et al., "Traditional and Complementary Medicine Use among Ebola Survivors in Sierra Leone: A Qualitative Exploratory Study of the Perspectives of Healthcare Workers Providing Care to Ebola Survivors," *BMC Complementary Medicine and Therapies* 20, no. 1 (2020), https://doi.org/10.1186/s12906-020-02931-6.

32. Temitope I. Borokini and Ibrahim O. Lawal, "Traditional Medicine Practices among the Yoruba People of Nigeria: A Historical Perspective," *Journal of Medicinal Plants Studies* 2, no. 6 (2014): 20–33.

33. Mohaddese Mahboubi, "Effectiveness of Myrtus Communis in the Treatment of Hemorrhoids," *Journal of Integrative Medicine* 15, no. 5 (2017): 351–58, https://doi.org/10.1016/S2095-4964(17)60340-6.

34. Abdullahi, "Trends and Challenges," 116.

35. Claire Munoz Parry, "Ebola: How a People's Science Helped End an Epidemic," *International Affairs* 93, no. 2 (March 1, 2017): 485–86, https://doi.org/10.1093/ia/iix043.

36. Charles M Good et al., "The Interface of Dual Systems of Health Care in the Developing World: Toward Health Policy Initiatives in Africa," *Social Science & Medicine, Part D: Medical Geography* 13, no. 3 (1979): 141–54.

37. MacLachlan, *Culture and Health*, 213.

38. Manguvo and Mafuvadze, "The Impact of Traditional and Religious Practices."

Chapter Four

A Critical Contextualized Approach to Studying Clashing Risk Cultures

Mapping the Transcultural Environmental Risk Communication of PM2.5 in China

HUILING DING AND JIANFEN CHEN

Over the past fifteen years, most of China's northern cities have been suffering from smog, especially in winter when coal is used for heating. In 2010 alone, outdoor air pollution was responsible for an estimated 3.3 million premature deaths all over the world, with most of the deaths occurring in low- and middle-income countries.[1] China alone contributed 1.2 million deaths to the total number. In 2012, when Beijing's smog issue became a global news headline due to a historical peak level of particulate matter (PM2.5) suspended in the air, China witnessed an unprecedented level of public attention to and engagement in smog reporting practices and advocacy for public policy changes.[2] As one of the main hazardous components in smog, PM2.5 refers to particulate matter with less than 2.5μm in aerodynamic diameter. A positive correlation has been established between PM2.5 concentration and respiratory diseases, asthma, myocardial infarction, and lung cancer.[3] The United States was the first country to set its National Ambient Air Quality Standards for PM2.5 in 1997, followed by the World Health Organization and other countries.

Despite its severe air pollution, China didn't incorporate PM2.5 into their national air quality standard until February 19, 2012, when China's Ministry of Ecology and Environment (MEE) announced that it would include PM2.5 in its newly released Ambient Air Quality Standards as one of the indices, taking effect in 2016. Such changes in China's air quality policy were introduced thanks to the tremendous public pressure exerted by social media movements that originated from a foreign agency and were pushed forward by celebrity opinion leaders and concerned nonprofit organizations and citizens.

To map out the route of risk communication of PM2.5 in China, this chapter takes a critical contextualized approach to global rhetoric of science by investigating how the risk of PM2.5 was communicated to and among the Chinese publics to generate public outcries that pushed for policy changes. It shows how Euro-American science, when practiced in emerging environmental risk communication in China, became a hybrid of transnational risk discourses and the bottom-up, participatory grassroots efforts made by local communities directly impacted by such risks. More specifically, we focus on the concerted efforts made by grassroots Chinese environmental organizations, news media such as *Southern Weekly*, celebrity figures, and publics to widely disseminate knowledge on the hazards of PM2.5 and call for public participation in air quality monitoring, efforts that eventually led to the change of China's air policy.

Our study reveals that this bottom-up movement to reshape China's air policy was ironically, though unintentionally, started by a foreign player, the US Embassy in Beijing (referred to as the US embassy in the rest of the chapter for the sake of brevity). Transmedia such as the cell phone app BeijingAir mediated the risk conception from the Euro-American actor to Chinese actors in this air-policy-change event. However, the successful civic engagement in risk communication of PM 2.5 and the ensuing air policy change came from the Chinese public's association of PM 2.5 with a health risk, which the government downplayed. The Chinese public leveraged Euro-American definitions of air quality (AQI) to navigate complicated socioeconomic and political systems in order to promote changes to air policy, reinforcing Huiling Ding's 2014 call for a critical contextualized method when investigating risks and uncertainties in non-Euro-American contexts.[4] This methodology, which employs six dimensions of inquiry—time-space axes, key players, tipping points, interaction analysis, power-knowledge relations, and contexts—provides a framework for understanding the complex and sometimes counterintuitive interaction of Euro-American and indigenous scientific rhetorics in transcultural communication about global events.

This chapter starts with a review of related works and studies on risk communication of PM2.5 as well as transcultural and transnational risk communication. After examining the traditional Euro-American rhetorical frameworks to risk communication and considering their inadequacy to fully capture the nuances of Chinese rhetorics of science in this risk event, we employ Ding's critical contextualized methodology to identify cultural key players that helped introduce a few "tipping points" in the PM2.5 event. We use the data collected from transmedia to explore how these key players took advantage of transmedia to exert geopolitical pressure on the Chinese government. Such transmedia tools and platforms coalesced into what Michel Callon, Pierre Lascoumes, and Yannick Barthe have called "hybrid forums" to initiate public deliberation about the severity of the PM2.5 situation in China and the possible environmental risks it could cause, to engage with institutional spokespersons, and to invite public participation in regional and national movements to measure air quality.[5] These *ad hoc* alliances eventually led to widespread public outcry and the subsequent policy changes at the regional and national levels to address the social-technical controversies of PM2.5 risk management.

Uncertainties, Spheres, and Hybrid Forums

The civic engagement in air policy change in China started with the controversy over the air quality standard. Understanding this controversy entails a review of the uncertainties that were its source. To better map out how different genres and communication products help mediate risk communication practice, we can draw on the theoretical insights and analytical tools offered by G. Thomas Goodnight, Lynda Walsh and Kenneth C. Walker, and Michel Callon, Pierre Lascoumes, and Yannick Barthe, adopting the spheres model of uncertainties and hybrid forums to examine the actors in the risk communication.[6] The spheres model treats uncertainties as "as creative rhetorical *topoi* (strategic stances or launching points) for inventing new discourses and new communities around shared risks."[7]

Goodnight introduced the framework of three spheres of argument in public deliberation—personal, technical, and public—which all function to "organize grounds of authority" and "serve as sites of controversy."[8]

The personal sphere only requires the use of "informal demands for evidence . . . and language use" to create disagreement and make arguments. The technical sphere is featured by "more limited rules of evidence, presentation and judgment" that have to be used by arguers of the field to achieve their communication goals.[9] The public sphere functions to handle "disagreements transcending private and technical disputes" with representative spokespersons using "common language, values, and reasoning" to settle disagreement about matters of interest to "the entire community."[10]

To describe how the socio-technical uncertainties get negotiated, Callon, Lascoumes, and Barthe introduced a highly productive theoretical concept of hybrid forums, which serves as "an apparatus of elucidation" that mediates the division that separates specialists and laypersons, or institutional spokespersons and average citizens.[11] For them, "Every hybrid forum is a new work site. It is a site for testing out forms of organization and procedure intended to facilitate cooperation between specialists and laypersons, but also for giving visibility and audibility to emergent groups that lack official spokespersons."[12]

Emphasizing "the multiplicity and complexity of uncertainties," Walsh and Walker furthered the understanding of Goodnight's spheres model of uncertainties and Callon, Lascoumes, and Barthe's conceptions on hybrid forums and called attention to "hybrid genres, forums, and communities that . . . reshape the boundaries of personal, technical, and public discourse." Walsh and Walker claim the spheres model provides "a rhetorical framework to slow down the hybridization of risk discourses so that scholars may examine the process and its political consequences." The hybrid forums "cross or combine standards of argument from different spheres" and thus allow "hybridization of arguments, genres, and communities."[13] In world risk contexts, hybrid forums represent sites that assemble a heterogeneous group of concerned stakeholders to explore political options for the reformation of technical, personal, and public boundaries.

Admittedly, the spheres model and hybrid forums are powerful frameworks and tools to study the "typical rhetorical effects of uncertainty argumentation in and across spheres."[14] Nevertheless, neither the spheres model nor the hybrid forums pay attention to the role of specific national cultural, socio-economic, and political factors, especially transcultural factors in risk communication across national borders. Because of this, we propose to incorporate these factors in the study of transcultural risk communication.

Incorporating Culture in the Study of World Risk Society

In his book *Risk Society*, Ulrich Beck asserted that, in today's globalized world, "poverty is hierarchic [and] smog is democratic" because "globalization tendency brings about afflictions, which are once again unspecific in their generality."[15] The key question in a risk society according to Beck is, "How can the risks and hazards systematically produced as part of modernization be prevented, minimized, dramatized, or channeled?"[16] With the boundary between risk and the cultural perception of risk becoming increasingly blurry, Beck described the "clash of risk cultures" or the "collision of culturally different risk realities" as "a fundamental problem of global politics in the twenty-first century."[17] Alan Irwin posited that "understanding the nature of risks and uncertainty is an important part of the scientific understanding needed both for many public policy issues and for everyday decisions in citizens' personal lives."[18] Examining "ecological and technological questions of risk and their sociological and political implications," Beck stressed that "questions have to be remembered, reposed, reconsidered, and rediscussed in a transnational setting, even if nobody has the answers."[19]

In her study of transcultural risk communication and viral discourses of the H1N1 flu pandemic, Huiling Ding proposes a theoretical framework of transcultural risk communication, which pays close attention to little-considered issues such as transnational connectiveness, flexible citizens, grassroots interventions, cultural contexts and differences, and transcultural virtual communities for the study of global risk politics.[20] In her analysis, she relies on Appadurai's notion of "grassroots globalization," or "globalization from below," which highlights the increasingly important roles that nongovernmental organizations (NGOs) play in mobilizing local, national, and regional groups on matters of equity, access, justice, and redistribution.[21] Despite the significant roles NGOs can play in grassroots globalization, little research has been done in risk communication to compare, describe, and theorize globalization from below and to systematically track its relationship to globalization from above (as represented by corporations, major multilateral agencies, policy experts, and national governments).

Researchers in China examined the tools and networks used by Chinese citizens to disseminate information and updates about air pollution. Kay, Zhao, and Sui found out that microblogging "helped precipitate the policy response from Chinese government" when China decided to include PM2.5 IN its air quality standard around 2013. They warned that, as a social media "unrepresentative of population in general and highly uneven along the lines of gender, class, and location," microblogging will not function as an "effective or just medium for public debate."[22] Focusing on civic participation in environmental governance during China's PM2.5 campaign in 2011, Fedorenko and Sun's mixed-methods study confirmed the important roles played by microblogging to mobilize laypersons to engage in environmental issues. They investigated civic participation in the environment in China, focusing on Shiyi Pan's microblogging as the battlefield of an air pollution campaign that mobilized millions of citizens and effected policy change.[23]

China's media censorship often prevents mainstream media, especially those who are government sponsored or controlled, from reporting issues that may contradict and confront existing policies or regulations. Because of these restrictions, risk communication in China often works bottom-up rather than top-down.[24] Ding's earlier work revealed that the initial outbreak of SARS in 2002 and 2003 saw scant official media coverage, but alternative media, including civic websites and word-of-mouth communication, were anything but silent.[25] Following the same vein, this study investigates both official media and alternative media to compare official and vernacular discourses that narrated the clashing risk cultures differently and to explore how two sociopolitical systems, environmental risk ecologies and complex cultures, interact with each other in negotiating possible solutions to sociotechnical problems.[26]

Method and Data Collection

As this study aims to map out the transcultural risk communication of PM2.5 in China, it relies on data from various media through the air policy change period from 2011 to 2012 to cover the key actors in this process. We borrow the term "transmedia" from Marsha Kinder, who first coined the term to describe the multiplatform and multimodal expansion of media content. [27] Since we trace the risk communication of PM 2.5 across various media platforms, "transmedia" serves as an overarching term to cover all such media platforms where the same issue is reported, discussed, described, presented, interpreted, debated, and more. Moreover, we employ the sphere model of uncertainties and critical contextualized methodology to analyze risk communication processes surrounding PM2.5 in China. Our research questions are as follows:

A Critical Contextualized Approach to Studying Clashing Risk Cultures | 93

1. How did the key actors in this process interact with each other to communicate the risk of PM2.5 to the Chinese public?
2. How did the participation of the Chinese public accelerate the change of China's air policy?

In order to cover the principal actors in this event, transmedia data have been collected from the following sources:

- the US embassy's webpage on air quality and PM2.5 readings published in 2011;
- Sina Weibo posts with PM2.5 as the keyword on Sina Weibo from October 22, 2011, to November 16, 2011, totaling 22,098 characters;
- Sina Weibo posts with PM2.5 as the keyword from November 16, 2011, to December 16, 2011, totaling 32,090 characters;
- the environmental nonprofit organization (ENGO) Green Beagle's Sina Weibo posts with PM2.5 as the keyword from July 1, 2011, to December 31, 2011, totaling 6,757 characters;
- seventeen articles published by the print media *Southern Weekly* from October 28, 2011 to June 6, 2012; and
- air policy documents published by the Chinese environmental authority ranging from 2011 to 2013.

As mentioned above, Ding's critical contextualized methodology employs six dimensions of inquiry: time-space axes, key players, tipping points, interaction analysis, power-knowledge relations, and contexts.[28] We defined our key players as international, national, regional, institutional, and communal forces that participated in the risk negotiations surrounding PM 2.5. Such key players include the following:

1. the US embassy, whose PM2.5 indexes circulated out of its intended American community living in Beijing to millions of Weibo users in China;
2. the ENGO Green Beagle that started the grassroots movement of PM2.5 monitoring and reporting;

3. two celebrity figures, Shiyi Pan (real estate) and Yuanjie Zheng (children's literature), who voluntarily assumed the roles of unofficial spokespersons for the concerned public;

4. the deputy director of the Beijing Municipal Bureau of Environmental Protection (BMBEP), Shaozhong Du, who engaged with Pan in Weibo exchanges; and

5. the *Southern Weekly*, a progressive newspaper based in Guangzhou that helped fuel the controversy with the symbolic power of the media.

Focusing on the PM2.5 policy changes, we limited our temporal axis to September 2011 to February 2012, when China's Ministry of Environmental Protection enacted the new and enforceable Ambient Air Quality Standards including PM2.5 (see table 4.1 for a list of key players, the timeline, and the types of arguments, i.e., technical, public, or personal, made by these key players). Our spatial mapping identified Beijing, Shanghai, and other tier-one cities as active participants in the movement. Focusing on risk monitoring and policy changes, our tipping-point analysis identified four key transformational moments that led to aggregated public action or intense negotiations through mass participation afforded by Weibo, namely:

1. Green Beagle's PM2.5 monitoring campaign;

2. Shiyi Pan's postings of PM2.5 indexes published by the US embassy;

3. Shiyi Pan's engagement with Shaozhong Du about the BMBEP's rationales for using a different air quality rating than the US embassy; and

4. the online polls initiated by Shiyi Pan and Yuanjie Zheng that attracted tens of thousands of participants and media attention.

What's interesting about these tipping points is their use of hybrid arguments that traverse the public, personal, and technical spheres to provide legitimacy to their arguments and to rally support from various stakeholder groups without alienating possible collaborators. Such strategies highlight the need to introduce grassroots movements and push for policy changes without alienating institutional players, a political dynamic

Table 4.1. Timeline for Risk Communication of PM2.5 and Hybrid Arguments

Timeline	Player	Activity	Arguments		
			Technical	Public	Personal
2008	US embassy in Beijing	Measuring PM2.5 using its own device		Yes	
Sept. 22, 2011	Green Beagle (ENGO)	Carrying out public monitoring of PM2.5 and Weibo posting AQI table of the US Environmental Protection Agency	Yes	Yes	Yes
Oct. 7, 2011	Green Beagle (ENGO)	Weibo posting harmful effects of PM2.5 and disagreement over China's air policy; forwarded over 3,000 times	Yes	Yes	Yes
Oct. 22, 2011	Shiyi Pan, a real estate celebrity	Weibo posting a screenshot of @BeijingAir from the US embassy; forwarded 4,496 times		Yes	Yes
Oct. 31 and Nov. 1, 2011	Shaozhong Du, deputy director of BMBEP	Weibo post explaining the difference between air quality ratings by the US embassy and BMBEP	Yes	Yes	Yes
Nov. 6, 2011	Shiyi Pan; Yuanjie Zheng, a renowned children's literature writer	Initiating online polls requesting MEE to set up enforceable standard for monitoring PM2.5 and surveying public opinion about Beijing air quality	Yes	Yes	Yes

continued on next page

Table 4.1. Continued.

Timeline	Player	Activity	Arguments		
			Technical	Public	Personal
Nov. 16, 2011	MEE	Releasing the second draft of Ambient Air Quality Standards, including PM2.5, effective nationwide in 2016	Yes	Yes	
Nov. 25, 2011	Green Beagle, five other ENGOs, and *Southern Weekly*	Submitting seven suggestions to MEE on the Ambient Air Quality Standards	Yes	Yes	
Feb. 29, 2012	MEE	Enacting new Ambient Air Quality Standards with PM2.5 included, enforceable in tier-one regions from 2012	Yes	Yes	

Source: Author provided.

that distinguishes the risk culture in China from those in Euro-American contexts. Power-knowledge and interaction analyses examine how these transnational, institutional, and extra-institutional players collaborated on the contested definitions of PM2.5 as a pollutant and its health impacts, rallied to help draw up inventories of PM2.5 readings, and resorted to existing technical arguments to help define the riskiness of PM 2.5. Finally, using contextual analysis, we investigate the political, economic, cultural, and material contexts in China that helped shape national environmental policies and risk responses from different cultural and communal players.

Findings

Environmental controversies are both fundamentally socioeconomic, with pollutants coming from industrial and residential sources, and sociotechnical, with experts and authorities collaboratively defining environmental risks and policies. What complicated the cultural politics surrounding PM2.5 in China was the national emphasis on economic performance, social stability, and nationality to maintain the political legitimacy of the Communist Party of China.[29] As a result, it was politically expedient for the Chinese government to subordinate the environment to capital and the free market. This official silence about the health risks of PM2.5, however, compromised multiple goals of citizenship, namely, rights to knowledge, information, and participation, the guarantee of informed consent, and the limitation of the total endangerment of collectivities and individuals.[30]

Our analysis identifies the grassroots use of hybrid forums, including Weibo posts by NGOs, celebrities, and institutional authorities, as well as online polls initiated by celebrities, as the primary mechanisms for introducing public deliberation and the subsequent institutional responses and recommended policy changes. Multiple key players participated in the vernacular negotiations about PM2.5 as a suspicious health risk that should be officially acknowledged, monitored, and regulated. Forming an ad hoc coalition, these actors actively participated in or, in the case of the US embassy, were passively recruited by domestic stakeholders into the deliberation processes. These multiple overlapping threads of negotiations helped to "set politics free by changing the boundaries of the political so that it becomes more open and susceptible to new linkages as well as capable of being negotiated and reshaped."[31]

The US Embassy in Beijing: PM2.5 Risk Beyond the Border

Since China didn't include PM2.5 in its air quality index until 2012, the US embassy began to use its own monitor to measure the air quality from 2008 due to its concern about the health of the US community in Beijing. These air quality ratings complied with the US National Ambient Air Quality Standards, which has incorporated PM2.5 since 1997.[32] To serve the US community in Beijing, the US embassy published hourly numerical updates of the AQI on two public platforms, its official Twitter account (@BeijingAir) and its website. As a foreign institution in China, the US embassy made cautious claims about its air quality monitor, which was only intended to be used as "a resource for the health of the American community." It also warned that "citywide analysis cannot be done based on data from [our] single machine."[33]

Close analysis of the US embassy's updates reveals its nature as a hybrid forum with its use of technical and public arguments. These updates included a three-column table using six different colors from green to crimson to represent the numerical readings of AQI, levels of health concerns, and meaning of those measurements (see figure 4.1). With the numerical value increasing from 0 to 500, the level of health concern moves from "good" to "hazardous" and the interpretation varies from "satisfactory air quality with little or no risk" to "serious health effects for everyone." These updates were designed for limited consumption by Americans in Beijing with both technical information of AQI readings and interpretation about possible health concerns.

With thick smog sweeping Beijing more frequently in 2011, Beijing citizens were anxious to know why the air quality was rated by the BMBEP as "qualified" or just "slightly polluted." While no one openly denied the possible negative health impacts from air pollution, little was known about the possible health consequences from PM2.5 pollution and the precise nature of health risks brought by exposure to the pollutants. Driven by

Air Quality Index Levels of Health Concern	Numerical Value	Meaning
Good	0 to 50	Air quality is considered satisfactory, and air pollution poses little or no risk
Moderate	51 to 100	Air quality is acceptable; however, for some pollutants there may be a moderate health concern for a very small number of people who are unusually sensitive to air pollution.
Unhealthy for Sensitive Groups	101 to 150	Members of sensitive groups may experience health effects. The general public is not likely to be affected.
Unhealthy	151 to 200	Everyone may begin to experience health effects; members of sensitive groups may experience more serious health effects.
Very Unhealthy	201 to 300	Health warnings of emergency conditions. The entire population is more likely to be affected.
Hazardous	301 to 500	Health alert: everyone may experience more serious health effects.

Figure 4.1. AQI levels and meanings released by the US embassy in Beijing. *Source*: United States Embassy in China, https://china.usembassy-china.org.cn/air-quality-monitor-2/?_ga=2.198681705.756809114.1657124461-262734213.1657124461.

the uncertainties surrounding Beijing's technical rationales for air quality ratings, a rhetorical opportunity was created for various stakeholders to collaborate, better evaluate, communicate about, and manage the perceived health risks. With Twitter available in the Apple store for free, the PM2.5 updates from the US embassy were quickly circulated out of their intended small community to a much larger ecology of concerned Chinese organizations and citizens.

Green Beagle and *Southern Weekly*: "I Gauge the Air Quality for My Motherland"

Appadurai identified NGOs as a stakeholder group that can play critical roles in grassroots globalization to introduce bottom-up policy changes.[34] Our analysis identifies a leading ENGO in China, Green Beagle, as one such player that helped mobilize mass participation in PM2.5 monitoring. Based in Beijing, Green Beagle started to track PM2.5 using its own devices starting on July 7, 2011. On September 22, 2011, Green Beagle initiated a public activity to call for participation in the global "No Car Day," driven by increasing public concerns over severe air pollution and uncertainties surrounding health hazards caused by the air problem.[35] Its volunteers walked from the Second Ring Road to the Fifth Ring Road of Beijing to measure PM2.5 levels in the city.

Green Beagle published six posts on PM2.5 testing results shared by its members and volunteers at varied locations in Beijing. These posts were accompanied by photographs showing volunteers using devices to monitor air quality in various urban areas, such as a community park (0.015mg/m3), overpasses with heavy traffic (0.032mg/m3), next to specific vehicle models (0.030mg/m3 at a parked SCR110 Honda motorcycle), and in smoking areas, to examine how levels of PM2.5 were affected by different possible variables. It also published a post sharing the US embassy's AQI table in English and encouraging public use of the table to better understand the health impacts of varied levels of PM2.5. Meanwhile, the same post explained that PM2.5 was the chief culprit of the smoggy weather in northern China, which mainly came from vehicle emissions.

Green Beagle took advantage of its technical knowledge about PM2.5 and chose the public forum of Weibo posts to disseminate technical data gathered from its members' air quality testing activities. Functioning as the intermediary between technical specialists and laypersons, Green

Beagle assumed the role of self-appointed spokesperson to advocate for people's right to be informed of both the levels of PM2.5 in Beijing and the possible harm caused by it.

About two weeks after its September activity, on October 7, 2011, Green Beagle published a post about the harmful effect of PM2.5 and the disagreement over China's air quality standard. Forwarded over three thousand times with over ninety comments, this post highlighted one important outcome of not including PM2.5 in China's air standard: Beijing's air was rated as "good" for 78 percent of 2010, even though on those "good" days people could see and smell the gray sky and smoky air. It also argued that if China had followed the international standard for air pollution, over 80 percent of those days would have been rated as polluted. Therefore, the so-called "good air quality" days were created by China's own standard, which purposefully excluded any data about PM2.5.[36]

Green Beagle's impactful public activity of monitoring the air quality caught the attention of *Southern Weekly*, which published a special report titled "I Gauge the Air Quality for My Motherland" (see figure 4.2) on October 28, 2011. The cartoon in the article said, "The prolonged delay of including the PM2.5 index in the national monitoring system resulted

Figure 4.2. Cartoon: "I Gauge the Air Quality for My Motherland," published by the *Southern Weekly* on October 28, 2011. Source: *Southern Weekly*, October 28, 2011, https://www.infzm.com/contents/64281.

in a national wave of self-monitoring and self-reporting activities among Chinese citizens. Like bamboo shoots showing up after spring rain, such grassroots participation has been pressing for official responses."[37]

This report attracted nationwide attention and was reprinted and circulated by several major news outlets, including *NetEase News* and *Tencent News*. Frustrated by the repeated official refusal to offer PM2.5 monitoring and anxiety about health risks, Chinese citizens started to use their own devices to collectively figure out the real picture of the air quality, which amounted to what Callon, Lascoumes, and Barthe conceptualize as a "collection of cases" to evaluate health risks caused by PM2.5.[38]

Celebrity Shiyi Pan: Launching a Weibo Campaign about PM2.5 in Beijing

As the managing director of SOHO China, the largest prime office real estate developer, Shiyi Pan is known not only for his wealth as a real estate celebrity but also for his innovative way of creating a positive, hybrid public image. As a writer and an actor, he has published ten books and acted in three movies. Being one of the early adopters of digital media, Pan maintained an active presence on Weibo and attracted over seven million followers in 2011.

In mid-October 2011, the thick smog in Beijing forced the city to temporarily close several highways and airports. On October 22, 2011, Pan published three Weibo posts containing three different screenshots of Beijing's air quality results released by the US embassy (see figure 4.3).[39] In one screenshot, Pan added the exclamation "Oh my Gosh! Toxic" as orange text on top of the original image.

Widespread public attention was attracted by the stark difference between the "toxic" air shown on this screenshot and the "good" or "slightly polluted" air quality released and rated by the BMBEP. The post also aroused public interest in finding out the truth about the air quality and the causes for the discrepancies between official data and the conflicting data from the US embassy. This screenshot post was forwarded 4,496 times with 1,568 comments. With Pan's burst of Weibo posts, the concept of PM2.5 moved out of tweets posted by the US embassy and circulated widely in the public sphere of Weibo. Pan published a follow-up Weibo post about the source of his screenshot, a free Apple app called BeijingAir. Pan's screenshot and the app BeijingAir provided an alternative yet easy

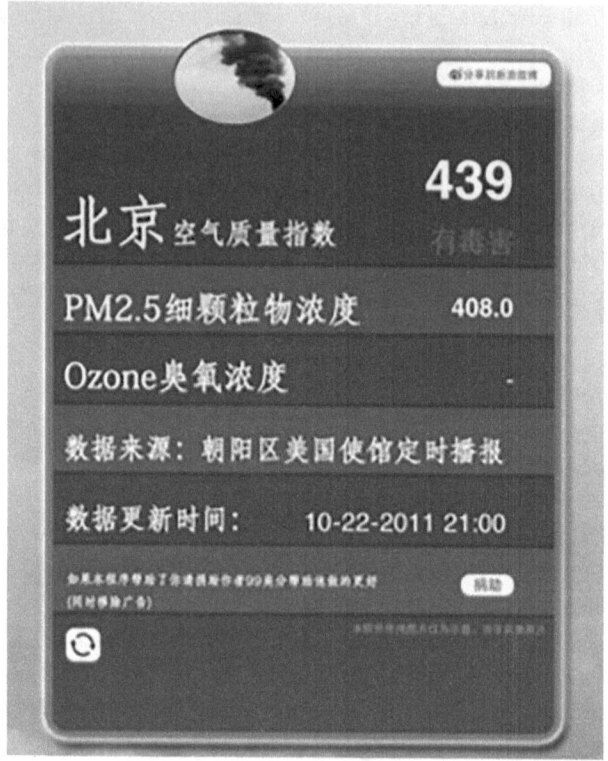

Figure 4.3. Screenshot of Pan's Weibo Post on October 22, 2011. Translation: "Beijing Air Quality Index 439 toxic; PM2.5 concentration 408.0 Ozone concentration." *Source*: US embassy in Chaoyang District.

way for the public to access and compare the different air quality ratings by the US embassy and BMBEP. Thus, he pushed himself to the center of the public debate on the truth of Beijing's air quality. On November 3, Pan published a post that went viral: "The BMBEP should release accurate, comprehensive, and real air quality data. Only by understanding the seriousness of the problem can people solve this problem with shared goals and concerted efforts. Poor air quality was a significant concern for all parties involved instead of only the Environmental Bureau or individual leaders. Don't get us wrong—we are not working to create media hype or to fight against the Environmental Bureau."[40]

Relying on arguments about the public right to know and emphasizing his goodwill, Pan carefully constructed his position as one focusing on access to technical evidence instead of one defying authorities or generating public outcries. Pan's Weibo activities triggered outrage among netizens and incited public backlash against the way PM2.5 had been excluded in Beijing's AQI data. His post was forwarded 2,861 times and attracted a total of 1,279 comments, with most of them supporting Pan's call for the official release of the PM2.5 index.

BMBEP's Shaozhong Du: An Ambiguous Response to Public Concerns about BMBEP's Air Quality Data

As the deputy director of BMBEP, Shaozhong Du was one of a few Chinese officials who used Weibo to respond to and communicate with the public. As an institutional spokesperson, Du employed hybrid arguments that resorted to his technical expertise to address and alleviate public concerns in his online debate with Shiyi Pan on Beijing's air quality standard.

After the massive engagement with Pan's posts about the US embassy's air quality index, Du published Weibo posts on October 31, 2011, to respond to Pan's posts and the subsequent public concerns about the discrepancy between the air quality ratings provided by BMBEP and the US embassy. He attributed the different results to the technical data used by the US embassy and the BMBEP, since the former used the median value and the latter used the average value.[41] He also emphasized that the air quality data issued by the US embassy was intended for internal use only and questioned the "scientific rigor" of the embassy's data release methods, which to him looked more like hype than objective research. His ambiguous responses only resulted in more public suspicions and questioning of the BMBEP's air quality ratings. On November 1, 2011, Du admitted in a Weibo post: "We've been too weak and too slow in communicating to the public the measurement of ambient air; at the same time, the environmental information is far from understandable. Because of this, concerned citizens are exposed to hard-to-understand technical terms such as standard, regulation, concentration, index, annual average, and daily average. Moreover, the public was further confused by the data issued by a foreign embassy, the consecutive days of fog, and the online discussion about air quality standards."[42] To prove to the public the validity and accuracy of BMBEP's air quality ratings, he invited Pan, Yuanjie Zheng, and other public figures to visit BMBEP and check the equipment for air quality measurement. In his later posts, he compared the number of days of good air quality in Beijing in 2010 with those in 1998, arguing that Beijing's air quality had improved immensely over the past decade.

Nevertheless, the public didn't buy his argument and insisted on knowing why BMBEP didn't make PM2.5 known to them. Instead of directly addressing public concerns, his post advocated that more attention should be paid to emission reduction. As the institutional spokesperson, Du succeeded to a limited degree in connecting with the public and showing them how the facility worked. His ambiguous and defensive messages, however, failed to engage fully in public debates about the quality or accuracy of BMBEP's data or to alleviate public distrust of BMBEP's official data or technical criteria.

Celebrities, Online Polls, and *Southern Weekly*: Collecting the Cases of Public Concerns

At the peak of his involvement in this public event, Pan initiated an online Weibo poll on November 6, 2011, surveying the public interest in setting up both an enforceable air quality standard to monitor PM2.5 and a timeline to formulate such standards. He justified his poll by focusing on the critical role policy updates play in ensuring compliance from organizations and individuals:

> According to experts, PM2.5 in the air is extremely hazardous to human health. Only after the State releases mandatory standards on the measurement of air quality can every city carry it out. Only with an understanding of the severity of air pollution will everyone take conscientious action to reduce air pollution and to change unhealthy lifestyles. Please participate in the poll and share this post. One week later, I will write a letter to present the result of this poll to the Minister of CMEP.[43]

Pan's poll attracted 42,188 respondents, with 91.1 percent stating that they would like to see the air quality standard come into effect in the same year.[44]

Yuanjie Zheng, a renowned Chinese writer in children's literature, was another celebrity who played an active role, using his Weibo to push for PM2.5 policy changes. As the second most active individual following Shiyi Pan, Zheng published fourteen Weibo posts, which were forwarded over a thousand times between October 22 and November 16, 2011. Rejecting BMBEP's claim about Beijing's improved air quality, Zheng launched an online poll on November 6, 2011, the same day that Pan initiated his online poll. Inviting the public to express their opinion about the air quality of Beijing, Zheng asked participants to choose from three positions: did they believed Beijing's air quality had improved, worsened, or remained unchanged? Taken by 6,993 people, Zheng's poll revealed that 88.8 percent of participants believed that the air quality was getting worse.[45]

The social media platform Weibo functioned as a hybrid forum that provided both Pan and Zheng the opportunity to test various ways to organize virtual grassroots campaigns and to function as self-appointed spokespersons who gave visibility and audibility to widespread public concerns about PM2.5. These two polls served as hybrid arguments that publicized technical knowledge about PM2.5 and its health impacts, collected inventories, and assessed public opinions. Their results revealed the public outcry for transparent risk information and quick adoption of internationally accepted risk assessment measures.

On November 23, 2011, *Southern Weekly* pushed the civic campaign of air quality policy changes to a new level by forming an ad hoc alliance with six of China's leading ENGOs, including Green Beagle. Working with these groups, *Southern Weekly* formulated seven suggestions to the MEE for the second draft of the Ambient Air Quality Standards, which was both mailed to the MEE and published in the newspaper.[46] The suggestions included accelerating the timeline of the enforcement of revised standards, using colors in the Air Quality Index to distinguish and better communicate the severity level of air pollution, and offering health-alert messages to special regions and populations who are more sensitive to AQI changes. This collaborative effort was significant because it represented the cumulative results from both the hybridization of arguments and communities and collaborative political efforts, combining technical, public, and personal arguments.

Changes in Regional and National Air Quality Policies

Our analysis so far reveals the positive impacts exerted by the hybrid arguments made by ENGOs and public celebrities on Weibo, the massive dissemination of these arguments through social media due to public concern, and the intensive coverage from *Southern Weekly*. The civic campaign of grassroots air quality monitoring supported the validity and relevance of the AQI data released by the US embassy. Meanwhile, environmental protection agencies found themselves pushed to the center of the PM2.5 controversy due to the debate on air quality standards between public celebrities and institutional spokespersons.

On November 13, 2011, seven days after Pan's and Zheng's online polls, a chief engineer of the Shanghai Environment Monitoring Center announced that Shanghai planned to include PM2.5 in its updated air quality monitoring standard.[47] On November 16, 2011, MEE released a new draft of its Ambient Air Quality Standards for public feedback, listing PM2.5 as one of the monitored indices. On February 29, 2012, MEE and China's General Administration of Quality Supervision, Inspection and Quarantine enacted the new Ambient Air Quality Standards, which included PM2.5 as one of the newly added AQIs. Although the new standard would not be in force across the nation until 2016, it set up the agenda for different regions to implement it. For example, on February 29, 2012, MEE announced that the tier-one regions in China, including Beijing, Tianjin, the Pearl River Delta, and the Yangtze River Delta, would start enforcing this standard in 2012. At the national level, in November 2012, the 18th National Congress of the Chinese Communist Party declared the national plan to develop China as an "ecological civilization" to enhance humanity-nature harmony under the epitome "Beautiful China."[48]

Studying Uncertainties in China: Call for Critical Contextualization

Our story about PM2.5 policy change in China seems complete until one digs deeper into the cultural, political, infrastructural, and socioeconomic contexts surrounding PM2.5 reform in China, with the last two components, power-knowledge interaction and context, offered by Ding's critical contextualized methodology. Only by looking more closely at the transcultural dynamics can one start to understand the material restraints and consequences of such changes. Though tremendously helpful in exploring hybrid forums and hybrid arguments in Euro-American contexts, the theories offered by Walsh and Walker and Callon, Lascoumes, and Barthe[49] have limited explanatory power when applied to non-Euro-American contexts, such as China, with very different material and socioeconomic conditions.

Because of this, we call for more nuanced frameworks to consider factors such as socioeconomic, cultural, and political settings, affordances, and constraints, especially power-knowledge interactions and changes along the temporal axis in risk communication in non-Euro-American contexts.

Our analysis of the larger socioeconomic contexts reveals some issues overlooked by our analysis of hybrid spheres and hybrid arguments surrounding the PM2.5 incident. In addition to technoscientific factors, numerous unique economic and political factors, as well as cultural and national values, contribute to such omission. Here we will focus only on the following four factors, which include:

1. socioeconomic and infrastructural constraints;

2. the Chinese Communist Party's emphasis on economic development and stability;

3. the emphasis of nationalism and the collectivist cultural tendency, which prioritizes national interest over individual needs and avoids direct confrontation with authorities to show goodwill toward the audience; and

4. the emerging field of technical and scientific communication in China, which helps connect specialists with the lay audience.

As a transnational player, the US embassy helped bridge the epistemic gap, though unintentionally, between the technical sphere and the personal sphere by providing an alternative interpretation of the PM2.5 situation in Beijing. However, as a foreign actor, the US embassy was seen as a potential threat to China's state security as Du questioned the scientific validity, rigor, and reliability of its PM2.5 ratings.[50] The US embassy was aware of the sensitivity surrounding its involvement in environmental controversies and refrained from such domestic disputes by posting the disclaimer that its ratings of PM2.5 were only for the health of the American community. Nevertheless, as a transcultural player, the US embassy served unintentionally as a catalyst to kick off the debate over the validity of PM2.5 published by BMBEP.

Considering the complicated nature of the PM2.5 controversy, hybrid forums such as Weibo offer individuals a platform to call attention to the disturbing air quality ratings from foreign sources such as the US embassy and to rally the netizens to voice their concerns. Notably, the public outcry was stirred up by celebrities who have the social influence to mobilize meaningful policy changes.[51] As celebrities, Pan and Zheng played important roles in collecting and articulating public concerns over PM2.5 ratings using the hybrid forum of Weibo. Their "depoliticized" tactics helped avoid direct confrontation with the authorities.[52] By associating air pollution risks with the shared goals of a greener home and public health, Pan and Zheng emphasized their good intention of promoting the health and happiness of all Chinese citizens and redirected public efforts to possible solutions at regional and national levels without highlighting political or ideological differences between the US and China.

Similarly, by avoiding direct opposition to the Chinese Communist Party or the state as environment activists, the ENGOs developed informal networks via hybrid forums such as Weibo and engaged in a form of "negotiated symbiosis" with the state.[53] They leveraged their technical expertise to publicize the knowledge of PM2.5 by engaging environmental enthusiasts in their activities. With both official and professional communication of PM2.5 and official ratings of PM2.5 missing in China, ENGOs played the role of professional communicators by advocating for the public's right to know the truth about air pollution. Meanwhile, using their own devices, they engaged the public in monitoring PM2.5 to explore alternative solutions to air pollution, a tactic that functioned as a symbolic gesture of their dissatisfaction with the official air quality standard. Thanks to the hybrid forum, their volunteering activities were captured by *Southern Weekly*, which published the cartoon titled "I Gauge the Air Quality for My Motherland." This humorous cartoon served as an outlet for the overwhelmed and frustrated public to express their concerns over air pollution issues. The synergy created by the ENGOs, celebrities, the general public, print media, and the foreign player of the US embassy helped push forward positive changes in China's air policy.

This air policy change was achieved at a critical moment when Chinese authorities began to realize the need to change China's economic growth mode to a more sustainable one. The previous high-growth and high-energy consumption mode had brought huge irreversible side effects to the country and its citizens. It was time for China to acknowledge the limitation of the previous growth mode and tackle its consequential by-product of air pollution.

Conclusion

This chapter proves how Walsh and Walker's spheres model of uncertainties and Callon, Lascoumes, and Barthe's hybrid forums helped us trace arguments and uncertainties across spheres in indigenous risk communication about emerging environmental risks. It also demonstrates that Ding's critical contextualized methodology can be helpful to investigate multilevel, multidirectional risk communication processes about PM2.5 among national, institutional, professional, communal, and key players via a wide range of media platforms.[54] Doing so allows us to plot the contour of uncertainties surrounding China's PM2.5 policies and highlights the need to investigate culturally specific material conditions, transmedia ecologies, socioeconomic contexts, political practices, and cultural values when examining clashing risk cultures in different countries. Only with culturally sensitive approaches can we reach some useful understanding of clashing risk cultures and possible ways to reconsider and reconcile them in a transnational setting.

Notes

1. Stephen S. Lim et al., "A Comparative Risk Assessment of Burden of Disease and Injury Attributable to 67 Risk Factors and Risk Factor Clusters in 21 Regions, 1990–2010: A Systematic Analysis for the Global Burden of Disease Study 2010," *The Lancet* 380, no. 9859 (2012): 2224–60.

2. Si-ming Lu, "A Case Study of Risk Communication: The Beijing Smog: The Communication Battle Between the Public and Government," *DEStech Transactions on Social Science, Education, and Human Science* (2016), https://doi.org/10.12783/dtssehs/emass2016/6804.

3. Xiaobei Deng et al., "PM2. 5-Induced Oxidative Stress Triggers Autophagy in Human Lung Epithelial A549 cells," *Toxicology in Vitro* 27, no. 6 (2013): 1762–70; Christopher H. Goss et al., "Effect of Ambient Air Pollution on Pulmonary Exacerbations and Lung Function in Cystic Fibrosis," *American Journal of Respiratory and Critical Care Medicine* 169, no. 7 (2004): 816–21; Annette Peters et al., "Increased Particulate Air Pollution and the Triggering of Myocardial Infarction," *Circulation* 103, no. 23 (2001): 2810–15; James C. Slaughter et al., "Effects of Ambient air Pollution on Symptom Severity and Medication use in Children with Asthma," *Annals of Allergy, Asthma & Immunology* 91, no. 4 (2003): 346–53.

4. Huiling Ding, *Rhetoric of a Global Epidemic: Transcultural Communication about SARS* (Carbondale: Southern Illinois University Press, 2014).

5. Michel Callon, Pierre Lascoumes, and Yannick Barthe, *Acting in an Uncertain World: An Essay on Technical Democracy* (Cambridge, MA: MIT Press, 2009).

6. Callon, Lascoumes, and Barthe, *Acting in an Uncertain World*; G. Thomas Goodnight, "The Personal, Technical, and Public Spheres of Argument: A Speculative Inquiry into the Art of Public Deliberation," *The Journal of the American Forensic Association* 18, no. 4 (1982): 214–27; Lynda Walsh and Kenneth C. Walker, "Perspectives on Uncertainty for Technical Communication Scholars," *Technical Communication Quarterly* 25, no. 2 (2016): 71–86.

7. Walsh and Walker, "Perspectives on Uncertainty," 72.

8. G. Thomas Goodnight, "The Personal, Technical, and Public Spheres: A Note on 21St Century Critical Communication Inquiry," *Argumentation and Advocacy: Special Issue: Spheres of Argument: 30 Years of Goodnight's Influence* 48, no. 4 (2012): 260, https://doi.org/10.1080/00028533.2012.11821776.

9. G. Thomas Goodnight, "The Personal, Technical, and Public Spheres of Argument: A Speculative Inquiry into the Art of Public Deliberation." *Argumentation and Advocacy* 48, no. 4 (2012): 202.

10. Ibid.

11. Callon, Lascoumes, and Barthe, *Acting in an Uncertain World*, 35.

12. Callon, Lascoumes, and Barthe, *Acting in an Uncertain World*, 36.

13. Walsh and Walker, "Perspectives on Uncertainty," 72, 74.

14. Walsh and Walker, "Perspectives on Uncertainty," 76.

15. Ulrich Beck, *Risk Society: Towards a New Modernity* (London: Sage Publications, 1992), 36.

16. Beck, *Risk Society*, 19.

17. Ulrich Beck, *World Risk Society* (Malden, MA: Polity Press, 1999), 12.

18. Alan Irwin, *Citizen Science: A Study of People, Expertise and Sustainable development* (London: Routledge, 1995), 37.

19. Beck, *World Risk Society*, 8.

20. Huiling Ding, "Transcultural Risk Communication and Viral Discourses: Grassroots Movements to Manage Global Risks of H1N1 Flu Pandemic," *Technical Communication Quarterly* 22, no. 2 (2013): 126–49.

21. Arjun Appadurai, "Grassroots Globalization and the Research Imagination," *Public Culture* 12, no. 1 (2000): 1–19.

22. Samuel Kay, Bo Zhao, and Daniel Sui, "Can Social Media Clear the Air? A Case Study of the Air Pollution Problem in Chinese Cities," *The Professional Geographer* 67, no. 3 (2015): 361.

23. Irina Fedorenko and Yixian Sun, "Microblogging-Based Civic Participation on Environment in China: a Case Study of the PM2.5 Campaign," *VOLUNTAS: International Journal of Voluntary and Nonprofit Organizations* 27, no. 5 (2016): 2077–105.

24. Yanshuang Zhang, "Microblogging and Its Implications to Chinese Civil Society and the Urban Public Sphere: A Case Study of Sina Weibo" (PhD diss., University of Queensland, 2015).

25. Huiling Ding, "Rhetorics of Alternative Media in an Emerging Epidemic: SARS, Censorship, and Extra-Institutional Risk Communication," *Technical Communication Quarterly* 18, no. 4 (2009): 327–50.

26. Huiling Ding and Jingwen Zhang, "Imagining Health Risks: Fear, Fate, Death, and Family in Chinese and American Online Discussion Forums about HIV/AIDS," in *Imagining China: Rhetorics of Nationalism in an Age of Globalization*, ed. Stephen John Hartnett, Lisa B. Keränen, and Donovan Conley (Lansing: Michigan State University Press, 2017), 237.

27. Marsha Kinder, *Playing with Power in Movies, Television, and Video Games: From Muppet Babies to Teenage Mutant Ninja Turtles* (Berkeley: University of California Press, 1991).

28. Huiling Ding, "Rhetoric of a Global Epidemic: Intercultural and Intracultural Professional Communication about SARS" (PhD diss., Purdue University, 2007).

29. Baogang Guo, "Political Legitimacy and China's Transition," *Journal of Chinese Political Science* 8, no. 1–2 (2003): 1–25; Heike Holbig and Bruce Gilley, "Reclaiming Legitimacy in China," *Politics & Policy* 38, no. 3 (2010): 395–422; André Laliberté and Marc Lanteigne, "The Issue of Challenges to the Legitimacy of CCP Rule," in *The Chinese Party-State in the 21st Century: Adaptation and the Reinvention of Legitimacy*, ed. André Laliberté and Marc Lanteigne (London and New York: Routledge, 2008), 1–21.

30. Philip J. Frankenfeld, "Technological Citizenship: A Normative Framework for Risk Studies," *Science, Technology, & Human Values* 17, no. 4 (1992): 462–65.

31. Beck, *World Risk Society*, 40.

32. "Particulate Matter (PM2.5): Implementation of the 1997 National Ambient Air Quality Standards (NAAQS)," 2008, accessed November 21, 2020, https://wikileaks.org/wiki/CRS:_Particulate_Matter_(PM2.5):_Implementation_of_the_1997_National_Ambient_Air_Quality_Standards_(NAAQS),_November_26,_2008.

33. "AQI Levels and Meanings Released by the U.S. Embassy in Beijing," *United States Embassy to China*, 2011, accessed November 21, 2020, https://china.usembassy-china.org.cn/embassy-consulates/beijing/air-quality-monitor/?_ga=2.149520370.559911029.1579965848-1378292718.1579965848.

34. Appadurai, "Grassroots Globalization and the Research Imagination."

35. Green Beagle, "No Car Day," Weibo, September 22, 2011, https://www.weibo.com/greenbeagle/profile?s=6cm7D0.

36. Green Beagle, "Blue Sky Created by Standards," Weibo, October 7, 2011, https://www.weibo.com/greenbeagle/profile?s=6cm7D0.

37. Jie Feng and Zongshu Lu, "I Gauge the Air Quality for My Motherland," *Southern Weekly*, October 28, 2011, http://www.infzm.com/content/64281.

38. Callon, Lascoumes, and Barthe, *Acting in an Uncertain World*, 22.

39. Shiyi Pan, "Paying Attention to Beijing," Weibo, October 22, 2011, https://www.weibo.com/panshiyi.

40. Shiyi Pan, "Interview with Luwei Luqiu," Weibo, November 3, 2011, https://www.weibo.com/panshiyi.

41. Shaozhong Du, "On the Differences of Air Quality Ratings," Weibo, October 31, 2011, https://www.weibo.com/dushaozhong?is_hot=1.

42. Shaozhong Du, "We are Weak in Communicating Air Quality," Weibo, November 1, 2011, https://www.weibo.com/dushaozhong?is_hot=1.

43. Shiyi Pan, "Online Poll Requesting CMEP to Set up As Soon As Possible Enforceable Standard on Monitoring PM2.5," Weibo, November 6, 2011, https://www.weibo.com/panshiyi. Note that CMEP is the acronym for the former name (Chinese Ministry for Environmental Protection) of the MEE.

44. Shiyi Pan, "Online Poll," https://www.weibo.com/panshiyi.

45. Yuanjie Zheng, "Online Survey for People's Opinion on Beijing's Air Quality," 2011, accessed November 21, 2020, https://www.weibo.com/zhyj.

46. Tao Wang, "Joining by 6 ENGOs, Southern Weekly Submitted Seven Suggestions to CMEP on the Second Draft of Ambient Air Quality Standard (for Suggestions)," *Southern Weekly*, November 24, 2011, http://www.infzm.com/content/65151.

47. "Shanghai Expects to Release PM2.5 data Next Year," *Top News*, November 6, 2011, http://news.sina.com.cn/green/news/roll/2011-11-15/065523467091.shtml.

48. "Green Water and Blue Sky: Overview of the Ecological Civilization Efforts after the 18th People's Congress," November 11, 2013, http://www.gov.cn/jrzg/2013-11/11/content_2525087.htm.

49. Walsh and Walker, "Perspectives on Uncertainty"; Callon, Lascoumes, and Barthe, *Acting in an Uncertain World*.

50. Du, "On the Differences."

51. Hugo De Burgh and Rong Zeng, "Environment Correspondents in China in Their own Words: Their Perceptions of Their Role and the Possible Consequences of Their Journalism," *Journalism* 13, no. 8 (2012): 1015.

52. Peter Ho and Richard Edmonds, *China's Embedded Activism: Opportunities and Constraints of a Social Movement* (London: Routledge, 2007), 336.

53. Ho and Edmonds, *China's Embedded Activism*, 337.

54. Ding, "Transcultural Risk Communication and Viral Discourses," 129.

Chapter Five

Where Voyaging Ends
Social Cosmology on Rapa Nui

FRANCISCO NAHOE

If they think of us at all, the rest of the world identifies Rapa Nui primarily as the home of the enormous, late Neolithic statuary that we call moai. Their impressive monumentality and utterly unique aspect—set against the romanticized backdrop of extreme geographical seclusion—so absorbs the imagination as to draw visitors more into the afterlife of the island's physical culture than a sustained consideration of the once wayfaring people who built them (figure 5.1). Nevertheless, to ponder the purpose and function of these colossal carvings immediately propels us into the world behind the moai. There we search for what their creators intended and why the project would merit the expenditure of the costly resources that supported it over the arc of many generations.

This chapter, which falls under the trope of metaphor, treats Polynesian wayfinding, the carving of moai, and Rapa Nui archaeoastronomy as species of social cosmology and so intends to expand our resources for understanding what constitutes the rhetoric of science in a global context. In particular, our investigation asks what we may gain from thinking of moai as the sort of quasi-discourse capable of "inducing cooperation in beings that by nature respond to symbols."[1]

Figure 5.1. The moai quarry at Rano Raraku. *Source*: Photo by Josefina Nahoe.

Amid Splendid Scarcity

The remarkable achievements of the ancient Rapa Nui people, called matamuʻa by their modern descendants, developed in an area of not quite 164 square kilometers.[2] Lava flows from late Pleistocene eruptions conjoined two earlier volcanoes to form the triangular island.[3] Though forested at the time of human settlement,[4] grassy hills now alternate with cones, craters, and small calderas. Pitcairn Island, barely inhabited today by fewer than a hundred progeny of *HMS Bounty* mutineers, lies more than 1,800 km to the west. To the east, one sails 3,700 km to reach the South American continent. Its location just beyond the Tropic of Capricorn provides the island with pleasant and temperate weather year-round: wetter in the austral winter, relatively dry in the summer. Rocky shores dominate the periphery of the island, which has few sandy beaches. Unlike many Polynesian islands, Rapa Nui has no coral reef, though islanders have always managed to supplement their largely cultigen-based diet with fishing.[5]

The archaic term for the Island conceptualizes Rapa Nui as the navel (*te pito*) of the world (*o te henua*). Anatomically, the umbilicus lies at the center of the body, but it also marks the last point of attachment to the mother. Nor should we take the designation whimsically. The easternmost outpost of ancient voyagers, Rapa Nui is the most isolated place of human inhabitation in the world. We don't know exactly when the first people arrived, but scholarly views vary from roughly the 4th to the 13th century of the common era.[6] Linguistic and genetic indicators strongly suggest an archipelago in eastern Polynesia, very possibly the Marquesas Islands, as the point of origin. Did lost seafarers make a unique landfall[7] or did some period of back-and-forth sea travel take place?[8] One thing is clear. However long the period of voyaging may have lasted, the lack of canoe-building materials on our island brought the long-range navigational culture of Polynesia to a definitive end. Thus, *te pito o te henua* became the last point of contact with the seafaring civilization that gave it birth.

After their ocean-going vessels deteriorated, however, the Rapa Nui launched themselves into a new, but demonstrably related, project. They carved moai. Embedded now in the physical culture of the island, the matamuʻa took advantage of rhetorical strategies that once preserved seafaring knowledge, especially ethnoastronomy, in response to hardship on land. The obdurate terrain of Rapa Nui only grudgingly supports human habitation, but the newcomers survived, and even came to thrive, because they found a way to marshal human resources. No speeches argue in favor of this kind of communal effort among our ancestors; no debates reveal their attitudes toward the technology they deployed or its environmental impact; no libraries catalogue and cross-reference their findings. Their work, inscribed now into the natural landscape of our island, provides nonetheless a splendid and memorable text that I have called a *quasi-discourse*, since moai enlisted support for the configuration of a social cosmology that favors survival. By constructing immense stone statues and arranging them upon ceremonial platforms, some of which prominently mark the same solar, lunar, and sidereal events relevant to navigation, the post-voyaging generations of Polynesians on Rapa Nui undertook specific activities entirely commensurate with conceptions of rhetoric and science found elsewhere on the globe. By these means, the matamuʻa induced successful, long-term cooperation in food production and freshwater management in the face of endemic scarcity.

A Journey of No Return

Polynesian voyaging may well have had its origins in the rapid expansion of the Lapita, an Austronesian people, between 1600 and 400 BCE.[9] With no metal instruments nor even a written form of their language with which to log discoveries, ancient navigators still found every inhabitable island in the nearly 26 million square kilometers of ocean of the Polynesian Triangle.[10] Wayfinding, as practiced by the traditional *ho'okele*,[11] required the minute observation of natural phenomena, including superb olfaction, a prodigious and detailed visual memory, and the capacity to recite verbatim both the lengthy and specific memorates of many previous generations of navigators as well as all the other units of oral lore relevant to the Pacific and its islands. Their powers of retention and recall "would have made Quintilian stare and gasp."[12] The intergenerational transmission of this encyclopedic natural history endowed wayfinders with practical knowledge of winds and swells, wave patterns, currents, bird flight,[13] pelagic mammals and fishes, and the characteristically different smells of ocean water and atmosphere in either epipelagic or shore zones. They learned direction heading without a compass and used stick charts as mnemonic devices to aid the recognition of known island wind patterns, waves, and swells. Indeed, a navigator would have been able to distinguish multiple overlapping swells just by the rocking of the vessel.[14] Above all, the *ho'okele* knew the night sky. Subdividing the heavens into segments,[15] they kept watch upon the progression of planets, stars, and asterisms while others in the crew slept. Having no written charts, Polynesian navigators would have had to visualize internally every celestial position at all points of the journey in order to gauge the bearings of the canoe at any given stage.[16]

In an astonishing 1976 expedition, a Micronesian wayfinder, Mau Piailug (1932–2010), using only traditional Oceanic methods and ethnoastronomy, navigated the Hōkūle'a from Hawai'i to Tahiti in thirty-five days. Having no need of nautical charts, sextant, compass, chronometer, or almanac, the crew sailed the flagship of the Polynesian Voyaging Society from Honolua Bay of Maui across the open ocean to the harbor at Pape'ete. The stirring arrival of a traditional voyaging canoe in French Polynesia exhilarated the ethnicity throughout our many islands, Rapa Nui included. In the scientific community at large, admiration for the sheer audacity of the Hōkūle'a voyage launched a new wave of scholarship across the disciplines into the scope and practices of ancient Pacific navigation. Subsequent research in archaeoastronomy, archaeology, and genetics have, in turn, forcefully asserted the capacity of prehistoric Polynesian wayfinders not only to undertake and repeat the precise and systematic navigation of the vast oceanic highways of the Pacific, but even to discover hitherto unknown islands.

Might ancient navigators have happened across Rapa Nui and then gone on elsewhere? Did they return to other South Pacific settlements or forge onward to the continental mainland?[17] Did they eventually find their way back to our island and stay? That Neolithic Polynesians possessed sufficient navigational perspicacity regularly to have made such long Pacific voyages, of course, is no guarantee that in fact they did so. Although the scholarship generated in response to contemporary wayfinding demonstrations from Hawai'i certainly offers intriguing possibilities about the relationship of our island to the rest of precontact Polynesia, the physical evidence does not yet supplant the hypothesis that Rapa Nui culture developed primarily as an outlier lacking sustained contact and exchange with contemporaneous human societies.

Despite the notable recovery of ancient wayfinding techniques and the practical illustrations of their efficacy undertaken by the Polynesian Voyaging Society,[18] a precise reconstruction of early Rapa Nui settlement has yet to emerge. Nor can we answer completely the question of whether and how ongoing contact with other islands might have taken place. Rapa Nui might have served as the place of embarkation from which Polynesian navigators found their way to the Americas. Though somewhat unlikely, the matter is not settled. Current research in Pacific archaeobotany,[19] for example, concludes that non-human mechanisms transmitted the sweet potato (*Ipomoea batatas*), or *kumara*,[20] throughout the many islands. The question of how *manioca* (*Manihot esculenta*) came to Oceania, however, remains uncertain.[21] Was Rapa Nui a waystation in its transmission from South America to the rest of the Pacific? More promising research comes in the form of recent human genetic studies,[22] which find the presence of Amerindian haplotypes in skeletal remains of precontact Rapa Nui. Since, *pace* Heyerdahl,[23] we have no convincing corroboration of long-range seafaring technologies on the part of the pre-Columbian populations on the continent, parsimony seems to favor the hypothesis of Polynesian navigation to and from the Americas.[24]

For many centuries, ancient Polynesians undertook long and dangerous voyages across the open ocean. But why? Firstly, island life at the site of any given settlement was always precarious. Throughout the Pacific, environmental stressors generated by late Holocene climate change,[25] volcanic eruptions, or seismic activity could potentially have impacted food production and population stability in any generation, especially in eastern Polynesia. Scarcity, in turn, can easily give rise to internecine conflict. We should exercise caution, however, in trying to assess the motivations of these intrepid voyagers solely on the basis of our own contemporary criteria. In addition to deep sea fishing, raiding, conquest and trade, all of which could well derive from need, David Lewis (1917–2002) takes seriously the stimuli of adventure and the pride of the navigators themselves in their craft and skill.[26] Citing a Marquesan paradigm documented early in the 19th century,[27] Lewis further underscores a well-known impulse, instigated by religious or visionary figures, that he describes as a "journey of no return deliberately undertaken toward some mythical or very ill-defined destination."[28] Any of these considerations, none of which necessarily excludes the others, might have induced emigration, especially in a social context that preserves star-guided wayfinding, practices interisland voyaging, and believes itself capable of surviving prolonged exposure to the high seas.

The Material and Spiritual Dangers of Landfall

The well-known narrative of Hotu Matuʻa, the legendary *ariki* (chief) and founder of Rapa Nui civilization, aptly illustrates each of these features. Collected by various ethnographers and enshrined in Rapa Nui songs still taught to children, this saga begins in the land of Hiva, a toponym common to the Marquesas archipelago.[29] There, Hotu Matuʻa ruled Maʻori from his home at Marae Reŋa. His wife, Vakai a Heva, lived with him, but his sister, Ava Rei Pua, dwelt at nearby Marae Tohio. Although tsunami, acute loss of arable land, and human demise constrained the *ariki* to plan for his people's embarkation, the motif of discord either with Oroi, his brother, or on his account, occurs so prominently in some recitations of lore as to suggest a further motive compelling departure. Hotu Matuʻa entrusted the preparation of a reconnaissance party to his counselor, Haumaka. In a dream, Haumaka navigated across the open ocean to an island in the direction of the rising sun where he found the volcanic summits, freshwater sources, and harbors of Rapa Nui. When his spirit returned to Hiva, Haumaka awoke and exclaimed aloud his satisfaction. ʻIrā overheard his father's exclamation and inquired after its meaning. Haumaka told the young man he had discovered the future home of Hotu Matuʻa and the people of Maʻori. At once, he began to instruct his son in the particulars of travel, fishing, disembarkation, and planting. Then he sent 'Irā with another son, Rapareŋa, and the five sons of his brother, Huatava, to scout the new terrain.

Upon arrival, the seven sailed first to the islets, Motu Nui, Motu Iti, and Motu Kaukau, visible from the volcano Rano Kau. After a successful catch offshore, the scouting party struck the first fire to cook and eat. Then they encountered a spirit-turtle[30] that led them to the pink sand beach of Anakena, the most suitable place for Hotu Matu'a and the coming settlers to land. There, at Hiro Moko, the spirit-turtle came ashore. One by one, 'Irā, Rapareŋa, and the sons of Huatava tried to lift it but couldn't. The last to try was Ku'uku'u, who succeeded in moving the spirit-turtle but in doing so was struck by its flipper and mortally wounded. The others took him, dazed and blinded, to the interior of a nearby cave for shelter while the spirit-turtle returned to Hiva. The young voyager implored his companions not to leave him alone, so they erected cairns and instructed the stones to reply for them when the dying man called out for his brothers. Ku'uku'u died in peace, while the others, who brought moai with them, undertook their erection and ornamentation in preparation for the arrival of Hotu Matu'a and the other immigrants. Before the advance party could return to Hiva, however, the voyaging canoes of the *ariki* appeared on the horizon. From the summit of Rano Kau, the young men signaled to the king to warn him that island resources were sparse, but Hotu Matu'a answered simply that the homeland was also bad. He and his wife came in one canoe, but, ahead of the king, Ava Rei Pua and her husband, Tu'u Ko Ihu, approached shore in another. One variant of the arrival narrative, which specifies that the two vessels had been bound together until just before landing, seems to imply that at least one of them had been damaged or otherwise distressed. To prevent his sister from arriving before him, Hotu Matu'a extended his mana, or spiritual power, to slow down her canoe until he could overtake her. At landfall, the two women each gave birth on the beach—a boy and a girl.

Paymaster William Thomson (1841–1909) of the *USS Mohican* first brought a version of this settlement account to the ethnological record in 1889, but variants appear later in the work of English archaeologist Katherine Routledge (1866–1935), Bavarian friar Father Sebastian Englert, OFM Cap (1888–1969), Swiss ethnologist Alfred Métraux (1902–1963), and German epigraphist Thomas S. Barthel (1923–1997). Attempting to adduce the empirical elements of this or any ethnogeny may or may not be productive, but a general paucity of motifs concerning either sea travel or star-gazing characterizes the collected narrations and stands in sharp contrast to foundation legends from other Polynesian sources.[31] Although Haumaka's dream may be said perhaps to allude to the night sky, the account itself pays scant attention to ocean voyaging or star-based navigation, while it lavishes detail on sighting the island, surveying landing areas, planting, and fishing offshore. Mostly, however, the particulars focus our attention on the material and spiritual dangers of landfall itself, the memorialization of the dead (figure 5.2), the assertion that a kind of prototypical moai came to the island with the vanguard from Hiva, the challenges to productive agriculture, the immediate growth of the population, and the purposeful arrangement of stones *so that they speak*. On the whole, the Hotu Matu'a legend in all its variants strikes me as late in origin, certainly after canoe-building and long-range seafaring had ended and probably post-dating moai production, transportation, and erection as well.[32]

Figure 5.2. Ahu a Kivi, restored in 1960 by William Mulloy and Gonzalo Figueroa. *Source*: Photo by Josefina Nahoe.

Apart from petroglyphs of a double canoe, or *vaka moana*, discovered in 1975 at Oroŋo,[33] and uncertain intimations in surviving lore, we find very little to affirm a period of back-and-forth ocean voyaging after settlement. Needless to say, the ongoing construction and maintenance of seaworthy vessels capable of transporting settlers required tremendous specialization and investment of social capital on other Polynesian islands. To my knowledge, however, the ethnographic material of Rapa Nui preserves very few hints that such was the case on our island. The types of wood used elsewhere to build voyaging canoes include *koa kā* (*Acacia koa*), *'uru* (*Artocarpus altilis*),[34] *tutui* or *ti'a'iri* (*Aleurites moluccanus*), and, especially, *tamanu* (*Calophyllum inophyllum*).[35] In addition to their suitability for building double-hulled vessels,[36] such species also have multiple practical and pharmacological uses. Though ancient wayfarers brought these trees into the Pacific from Southeast Asia and subsequently introduced them throughout Polynesia,[37] we have no record that what Patrick Kirch has called "transported landscapes"[38] included canoe-building trees on precontact Rapa Nui. To be sure, our ethnogeny specifies that the sons of Haumaka and Huatava brought food plants with them, especially *kumara*, sugar cane or *toa* (*Saccharum officinarum*), banana or *maika* (*Musa sapientum*), yam or *'uhi* (*Dioscorea alata*), and taro (*Colocasia esculenta*). Still, we have no mention of any intent to grow new forests, nor much reference to trees of any kind, except for *mahute* and *makoi*.

Although they used *mahute* (*Broussonetia papyrifera*) for clothing, not for building *vaka moana*, the early settlers may well have brought *makoi* (*Thespesia populnea*), which inhabitants of other Pacific islands valued as a component of ship craft. Nonetheless, even at maximal growth under climatic and soil conditions not matched on Rapa Nui, the *makoi* trunk would never have been large enough to hew out the hull of a seaworthy vessel. If natural disaster or local warfare urgently propelled Hotu Matu'a and his people into the seas, it may be that they had no time to assemble other saplings. Alternatively, for any number of agricultural or social reasons, attempts to plant and cultivate timber from which to build voyaging canoes may simply have failed.

Excluding mention of cultigens, Rapa Nui ethnography conspicuously lacks the wealth of ethnobotanical references found elsewhere in Polynesia.[39] Hunt correctly describes island biota as "relatively depauperate" and, summarizing both Flenley and Orliac, notes that "many more woody plants once grew on Rapa Nui."[40] Nonetheless, Carl Skottsberg (1880–1963) concludes that "even if man is responsible for disappearance of part of the flora, it cannot [ever] have been rich."[41] Nor is the island topsoil especially fertile, as the early settlers so painfully discovered upon arrival.

Well after settlement, and probably after the period of moai construction, islanders used the wood of the toromiro bush (*Sophora toromiro*) for carving *kohau roŋoroŋo*, the Rapa Nui script. Possibly precontact in origin, these glyphs remain largely enigmatic, but likely functioned as aide-memoires for genealogical or calendrical recitations. If, indeed, they antedate contact with the West, they will certainly constitute the most remarkable instance of the independent development of writing on the planet. On the other hand, though they may not precede the first Europeans in 1722, the second group of visitors in 1770 do attest to some form of writing, though they do not mention *roŋoroŋo* tablets as such. Any form of writing at all would still represent a remarkable mimetic development within a period of less than fifty years. By 1868, however, the French missionaries could find no islander able to read the *roŋoroŋo*. In 1870, French gambler and former naval officer Jean-Baptiste Dutrou-Bornier (1834–1876) seized control of the island and ruled despotically until his violent death. A hundred years later, after accelerated environmental degradation brought on by his introduction of sheep grazing, the native toromiro would become extinct.[42] Indeed, the islanders themselves only barely survived.

Neither the *Paschalococos disperta*,[43] possibly Rapa Nui's only large tree, nor anything else would have been suitable for building new voyaging canoes.[44] If they had been, and voyaging beyond Rapa Nui had continued, it would then be difficult to account for the dearth of such indications in either the archaeological or ethnological record. Though once ubiquitous throughout the island, the Easter Island palm, a native cocoid species, disappears from the pollen record in the middle of the 17th century.[45] Scholarship widely concludes that human inhabitants caused the deforestation of the island, which has often led to the debilitating implication of culpability in popular accounts of our prehistory. While such simplistic fault-finding rightly provokes censure from Polynesian cultural activism,[46] the real issue centers on the questions of how and why the matamu'a brought down the palms. Is moai production implicated?

Preserving Consensus and Collaboration

The ubiquity of moai on Rapa Nui staggers the imagination. From their quarry to every part of the littoral, nearly a thousand lie scattered[47] over an island smaller than Nantucket.[48] Nor does their extreme scale amaze any less. At nearly 12 meters in height, the largest moai dwarfs the sarsens on Salisbury Plain and weighs in at more than 160 metric tons. The volcanic cone of Rano Raraku, birthplace of the vast majority of them, emerged as a vent of Poike in the late Quaternary eruptive activity that also produced Mauŋa Terevaka, the island's highest point at just over 500 m. While our ancestors carved most of the moai out of the lapilli tuff found throughout this sector, for some projects, they evidently preferred harder stone from other parts of the island. The basalt moai, Hoa Hakanānaʻia,[49] still unhappily captive in the British Museum, counts among the relatively few specimens not having their origin either in the caldera or its outer edge.

The word *ahu* describes at least 313 magnificent stone ceremonial platforms,[50] typically rectangular, upon which the transported moai were meant to stand. Sometimes enclosed in extraordinary facades of tightly fitted, interlocking basalt masonry, Rapa Nui ahu illustrate a remarkably high standard of lithic design, as indeed one would anticipate in a structure capable of supporting the massive tonnage of several moai. In the broadest sense, moai erected upon ahu represent the local augmentation to monumental proportions[51] of small wooden or stone carvings called *tiki* or *unu* placed at a low-lying ritual dais or paved area known elsewhere in eastern Polynesia as *marae* or, in Hawaiʻi, *heiau*. As we have seen, the Hotu Matuʻa narrative begins at a complex of *marae* in Hiva, the implied birthplace of moai. On Rapa Nui, extant ahu typically mark the ceremonial centers of particular *mata* or clans dispersed throughout the periphery of the island, but especially on the south coast.

What purpose do the ahu serve? We know very little of the phenomenology of ancient religion on the island. The relatively stable Polynesian pantheon familiar to other islands undergoes significant modification in our traditions. Because the spiritual figures of Mangareva, Pukapuka, or Tahiti perform different roles in Rapa Nui ethnographies, inferences from cognate sources do not necessarily answer the questions raised by the unique monumentality of our physical culture. Archaeological and archaeobotanical studies do, however, underscore the endemic trial of gathering and preserving freshwater, along with diverse obstacles to cultivating and producing food on the island. We should therefore ask how the moai carvers of Rapa Nui conceptualize themselves in light of their demanding environment. Viewed from this perspective, the locations of moai and ahu, together with the labor required to carve, transport, and erect them, appear as a kind of symbolic discourse aimed at organizing and reaffirming the social cosmology (figure 5.3). Production of moai anchors the strategies favoring survival in an explicitly ritualized context that elicits seasonal repetition and significant displays of social cohesion.

The matamuʻa persuade successive generations to maintain the social relationships that preserve life and promote resilience in an undersupplied island environment. Settlement expeditions of the kind that seek undiscovered islands require an outstanding commitment to communal solidarity. Everyone on the canoe or in the fleet depends upon the keen intelligence and sedulous resolve of the entire company, not only navigators, in order to survive protracted periods on salt water. But Polynesian wayfarers did not merely endure the rigors of the open sea—they willingly embraced them and flourished in doing so. Still, the limitless ocean clearly represents an exceptional state. How, then, do erstwhile seafarers preserve on bounded land the same high measure of consensus and collaboration that seafaring demands?

Figure 5.3. The ceremonial complex at Tahai, restored by William Mulloy in 1974.
Source: Photo by Josefina Nahoe.

The Mana of Matariki Rising

As we have already seen, ethnoastronomy in Polynesia plays an essential role in voyaging. For this reason, the presence of explicitly astronomical features at prominent ahu now suggests an intriguing path of investigation to address the question of how the matamuʻa elicit the same intense level of mutual cooperation in post-voyaging generations. Some ahu clearly serve as observatories. Archaeologists from the 1955 Norwegian Expedition postulate the solar orientation of Ahu Tepeu, an ahu at Oroŋo, and two ahu at Vinapū, one to the equinox and another to the summer solstice.[52] Later discovery of a perpendicular to the back wall of Ahu a Kivi was thought to indicate "the azimuth of the rising sun at the equinox."[53] In the 1990s, Harvard astronomer William Liller affirmed that more than twenty ahu mark solar events, including Ahu Raʻai, Ahu Toŋariki, and both ahu at Hekiʻi.[54] At the northernmost point of the island, the matamuʻa—perhaps alluding to their wayfinding ancestors—built Ahu a Taŋa in the shape of a canoe pointing north.[55] Nor did the ancients limit their interests only to the seasonal risings and settings of the sun. Ahu Huri a Ureŋa, for example, appears to mark either the solstice of the austral summer or the rising of the Matariki (Pleiades) asterism or both.[56] Alternatively, later investigation suggests that alleged solstitial and equinoctial orientations at the same ahu align better with asterisms, especially Matariki, Kete (Aldebaran and three other Tauri), and Taʻutoru (Orion's Belt).[57] Thus, Ahu a Kivi, previously thought to have been equinox-oriented, would rather face the heliacal setting of Taʻutoru, probably to mark a new lunar year at the following full moon.[58]

Throughout the island, one finds large stone towers with inner chambers called *tupa* and conical rock piles known as *pipihōreko*. The latter clearly mark territorial boundaries and sometimes appear at sites with *tupa*, as at Ahu Raʻai and other ceremonial locations. Ethnological accounts compiled by early visitors attribute fish and turtle sighting functions to the *tupa*, though Father Sebastián points out that at least some of them are too far from shore to serve as observation posts for fishermen.[59] If, however, the matamuʻa intended these structures not to look at the ocean, but to facilitate sighting the asterisms that coincide with the seasonal appearance of pelagic fishes and other sea fauna, then proximity to the shore is irrelevant. The matter has not been studied sufficiently to settle, but general indications so far suggest a decisive role for ethnoastronomy in defining and preserving the social framework.

On our island, the same Polynesian knowledge of the stars that figures so prominently in voyaging will thus find new forms of expression in the production and erection of moai and, perhaps, in the building and positioning of *tupa* and *pipihōreko* as well. By channeling a key element of navigation into a new social project, such constructions preserve awareness of the celestial events that come to signal the beginning of planting, harvesting, or deep-sea fishing. Though its precise contours are now lamentably opaque, social function nonetheless appears to conjoin archaeoastronomy with the physical culture of the island. More importantly, this conjunction creates an explicitly visual representation of the tribal compacts that distribute work to survive. The purposeful intersection of Matariki rising, or any such natural phenomena, with the collective labor of the matamuʻa thus displays mana.

Ecodynamics and Lethal Contact

While undeniably impressive, ahu and their moai, along with *tupa* and *pipihōreko*, on no account exhaust the archaeological wealth of Rapa Nui. The island constitutes the largest outdoor site of ancient culture in the Pacific and one of the more notable anywhere. Ride on horseback from the village to your campsite and even the least observant will mark again and again the prodigious topknots (*pūkao*) of moai, the foundations of prehistoric houses (*hare paeŋa*), stone enclosures for vegetable gardens (*manavai*), stone chicken coops (*hare moa*), chicken lodestones (*maʻea moa*), flat lava flows reworked for irrigation (*pāpā*), rainwater basins (*taheta*), thin slabs of worked basalt (*keho*), earthen ovens (*ʻumu tahu*), stone burial cists and crematory platforms (*avaŋa*), obsidian spear points (*mataʻa*), obsidian adzes (*toki*), an extensive network of roads (*ara*)[60] cut and filled for the transportation of moai and *pūkao*, paved sea ramps (*haŋa paeŋa*) for canoes (*vaka*), bas-relief rock carvings, lithic scatter here, petroglyphs there, and territorial spirits (*akuaku*) absolutely everywhere (figure 5.4).

Did carving, transporting, and erecting moai upon their ahu lead to deforestation and overpopulation? If large quantities of trees and many laborers were needed for these projects, then yes, they did. Still, those suppositions are by no means certain. Of the moai still in the quarry at Rano Raraku, so many remain frozen in the various phases of completion as to allow a precise reconstruction of the method and stages of sculpting and initiating transport: from embryonic tracings to first incisions; from

Figure 5.4. Ahu Vai Ure at Tahai with the outskirts of the modern village of Haŋa Roa in the background. *Source:* Photo by Josefina Nahoe.

medial chiseling to artful detail; from a process nearing its conclusion to a fully worked, gigantic figure ready for conveyance. The *ara moai*, that is, the roads of transit, are today scattered with broken megaliths, the casualties of poorly conceived transportation procedures or the imperfect execution of correct ones. In some instances, however, possibly in response to an initial outbreak of internecine warfare, the matamu'a simply abandoned the moai en route to their ahu. They lie there still.

Competing theories of how prehistoric laborers moved the moai abound. In 1955, Heyerdahl proposed simple dragged wooden sleds for supine transport. Mulloy worked out complex rocking wooden sleds with v-shaped wooden pivot legs for prone transport in 1960. Pavel offered a demonstration of side-to-side swiveling for upright transport in 1982. Love, in 1985, tried wooden runners on log rollers for upright, face-forward transport. Pavel and Heyerdahl together modified the side-to-side swiveling technique for upright transport in 1986. Pestun and Valeyev modeled something similar in 1988 for upright, forward-leaning transport. Van Tilburg proffered wooden sleds on log rollers for supine transport in 2005. In 2011, Hunt and Lipo likewise turned to side-to-side swiveling for upright, forward-leaning transport before the statue's foundation and center of gravity are finalized for erection upon the ahu.[61] Some of these techniques require extravagant use of wood, either for scaffolding or rollers or both. Others imagine teeming labor corps, which would require a large population. At least three demonstrations, however, suggest ways of transporting moai using a relatively small workforce and no wood: Pavel and Heyerdahl, Pestun and Valeyev, and Hunt and Lipo. These latter approaches also conform in a general sense to the ethnographic record, which insists that the statues walked from Rano Raraku to their present sites.

In 1982, at his hometown of Strakonice in southern Bohemia, Czech engineer Pavel Pavel replicated a 12-ton, 4.5-meter moai in concrete from photographs of originals and assembled a group of wildly enthusiastic local volunteers to move it. His efforts came to the attention of Thor Heyerdahl (1914–2002), who decided to test the idea in situ using two authentic moai.[62] The first, near Haŋa Roa Otai, stood at 2.8 meters in height and weighed roughly 5 tons. The other, at Toŋariki, weighed 9 tons and stood at 4 meters in height. At roughly the same time that Pavel began to work with Heyerdahl, two nautical engineers from Leningrad, Aleksandr Vitalyevich Pestun and Rustem Chingizovich Valeyev, proposed a similar procedure which they demonstrated with scaled models.[63] In 2011, Terry Hunt and Carl Lipo undertook a cognate method near Ka'a'awa on the windward side of O'ahu using a 5-ton moai replica, which they walked 90 meters in 40 minutes.[64] Notwithstanding criticism that their tack would expose moai to intolerable damage in transit, the walkers have offered a clear alternative to modes of conveyance requiring teeming labor corps and sleds, rollers, or platforms made of wood.

The matter of transportation is crucial to evaluating the ecodynamics of precontact Rapa Nui. In one scenario, our ancestors systematically replaced the native palm with their own cultigens to better provide for human nourishment and clothing.[65] In the alternative, they cut down the trees to convey moai to the ahu. The whole question of whether to praise the matamu'a for surviving the persistent insecurity of their harsh environment, or to blame them for preferring the works of their hands to the bounty of the land, hinges on the question of transportation.

Before the beginning of comprehensive restoration projects at the equinox-oriented Akivi-Vaiteka complex in 1960,[66] none of the moai remained standing upon their ahu. Though natural events or even poor engineering may have accounted for some of the ruination, mostly the moai were deliberately toppled. Although the ethnological record brims over with accounts of fierce conflict and recurrent brutality, we cannot ascertain the facts. Did the complex tribal structure of the island collapse under the weight of scarcity or did communal breakdown precipitated by other factors hinder food production? Again, the relationship of social complexity to the carving of moai and the cultivation of food is crucial to understanding what may have brought about the destruction of such costly monuments. At precisely this juncture, the outside world made lethal contact with Rapa Nui.

In 1722, Dutch commander, Jacob Roggeveen (1659–1729), brought three ships to the island on Easter Sunday. His landing party—134 men armed with muskets, pistols, and cutlass[67]—killed a dozen islanders, but reported standing statues, trees, and cultivation. They estimated a population of two thousand to three thousand, but could hardly have surveyed the entire island, including caves. In 1770, a brace of vessels under the command of Don Felipe González de Haedo (1714–1802) arrived from Peru. With a display of gunfire at landfall, the Spanish first annexed the island, and thereafter blithely forgot about it altogether. Their astonished reports of few women and children suggest to me that whole segments of the population had taken care on this occasion to conceal themselves from armed scouting parties. Again, the island's vast network of caves would have made it easy to elude outsiders. In 1774 and 1786, respectively, Captain James Cook (1728–1779) and Jean-François de Galaup, comte de La Pérouse (1741–1788) documented their impressions of the island. They found progressively fewer people and trees, less cultivation, more toppled statues. The explosive eruption of Tambora in 1815 and the successive years of ongoing geothermal disruption and radiative forcings must have devastated the fragile balance of food production on Rapa Nui, as indeed it triggered severe weather events, dropped the surface temperature of the planet, and produced famine across the globe. Frère Eugène Eyraud (1820–1868), a Sacrés-Cœurs missionary who came to the island in 1864, found no living trees and no moai standing upon their ahu. At the visit of the *HMS Topaze* four years later, a surgeon of the Royal Navy, John Linton Palmer (1824–1903), confirms the same. Though we can neither trace the diseases of contact, nor review in detail the precise etiology of social disruption, we do know that by 1889 a combination of colonization, livestock grazing, slave raids, and smallpox combined to reduce the native population to 111 persons, among them my great-grandparents.

Tropes of Environmental Disaster

As to what precipitated the toppling of moai and the desecration of ahu, scholarship falls into roughly two distinct camps. The first and most commonly known viewpoint theorizes anthropogenic environmental collapse: the Rapa Nui exhausted their natural and social resources in the obsessive production of monumental statuary and comprehensive violence ensued.[68] A newer thesis, however, imagines a highly stratified but generally stable society overrun at the time of contact by exposure to the various pathogens that Europeans carried. In my view, the vector of contagion would have been as much rhetorical as epidemiological, for the encounter with new and virulent technologies certainly dramatized to the matamu'a that their cosmos was no longer an island.[69]

After the moai themselves, the world knows Rapa Nui mostly through tropes of environmental disaster and endemic savagery. Indeed, some version of the collapse hypothesis appears so frequently and so prominently in contemporary literature as to suggest more about the anxieties of up-to-the-minute writers than persuasive command of difficult research across multiple disciplines. Though hardly the first to have invoked its rhetoric, William Mulloy (1917–1978) may have launched this trope into the popular view of Easter Island. In a short but masterful overview of island prehistory written for an audience of non-specialists,[70] he seized explicitly upon Paul Ehrlich's unnerving renewal of the Malthusian spectre.[71] While Mulloy addresses thorny interpretive tangles exposed by the emerging archaeological record known to him at the time, he frames them against the supposed backdrop of environmental degradation, resource depletion, and violent competition for food. Though he composed a cautious and monitory intervention, the tendency to attribute the demise of the matamu'a to their own willful spoliation of the island's ecosystem in order to construct their unhappy physical culture has since proven irresistible. Our present climate of global anxiety urges a hermeneutic of ecocide onto Rapa Nui. Even so, while asserting catastrophic non-sustainability, this tactic likewise displaces the threat of calamity to the smallest, most atypical location imaginable. In doing so, it grants the boon of catharsis with too little investment in pity and fear.

Borrowing from the field of evolutionary psychology, and especially Zahavi's handicap principle, archaeologists Hunt and Lipo view the physical culture of the island as exhibiting costly signals that symbolically aggregate and organize small clan units (*mata*) to facilitate the distribution of scarce food and water supplies that are always difficult to cultivate, collect, and manage. They note that "shared community labors such as construction of ahu and moai tend to be most prevalent when resources are sufficient to support group members yet are nowhere plentiful."[72] Monumental statuary consciously and deliberately projects a social order. Costly signaling, cultural elaboration, and bet-hedging are, in effect, the rhetoric of moai.

Carving, transporting, and erecting monumental statues generates social benefits by ritualizing communal labor. By persuading an isolated population to undertake such extraordinary activities, the moai favorably impact all collective tasks associated with the struggle to live and thrive in an always formidable environment. These statues embody the mana, or spiritual power, of our ancestors at least in this sense. Where voyaging once shaped island societies and aggregated human capital for the demanding tasks of wayfinding, settlement, fishing, farming, and animal husbandry, moai production would soon become the new symbolic discourse that preserved a recognizably Polynesian identity in the absence of canoes.

Where Voyaging Ends

The new wave of Easter Island scholarship thus posits an undaunted and creative population that magnified the visual tropes of community, formerly the *tiki* and *marae* of Hiva, to embody a social construct adapted for survival in a land of sparse flora with poor soil. Without the resources of rhetorical analysis, carving moai and sailing upon the ocean appear before us as essentially dissimilar activities utterly divorced from one another. This chapter has seen fit to efface that distinction by asking how visual metaphor on Rapa Nui exploits the science of sidereal navigation to construct a social cosmology capable of sustaining itself under harsh conditions.

In an environment characterized by scarcity, moai persuade the people to work collectively for clearly established and magnificently visible purposes. While early European visitors do appear to have encountered the physical culture of the matamu'a after its apogee,[73] the accelerated destruction of the ancient ceremonial structures and the drastic loss of population may both have been direct by-products of contact with the outside world. Collision with the West fragments the traditional symbolic discourse embodied in moai that had once been capable of holding together a complex but isolated society managing meager resources. Herein lies the real significance of Rapa Nui to global rhetorics of science in the present age. Like the ancient matamuʻa, none today may escape our changing island-planet. What magnificent works do we place in view that will persuade present and future generations to labor collectively for the benefit of our posterity?

The moai carvers of Rapa Nui conceptualized themselves as an integrated community, whole and entire, in relation to their island universe. Contrary to the popular notion that constructing the physical culture of the island depleted biological materials and ignored environmental concerns, the matamuʻa clearly grasped the perennial uncertainty of their existence at the last point of contact with the world. Unlike their ancestors from Hiva, however, they had no more canoes on which to depart. Survival, therefore, required the highest degree of social organization and cooperation. To establish and maintain themselves, the matamuʻa undertook an enduring rhetorical project in the construction, transportation, and erection of monumental statuary under the same stars that once guided them to new lands. Enormous themselves, moai stand against the enormous precarity of life where voyaging ends.

Notes

1. Kenneth Burke, *A Rhetoric of Motives* (New York: Prentice Hall, 1950), 46.

2. See Claudio Cristino, Patricia Vargas, and Roberto Izaurieta, *Atlas arqueólogica de Isla de Pascua* (Santiago: Editorial Universitaria, 1981).

3. Óscar González-Ferrán, et al., *Geología del complejo volcánico Isla de Pascua Rapa Nui* (Santiago: Centro de Estudios Volcanológicos, 2004).

4. See John R. Flenley, "The Palaeoecology of Rapa Nui, and its Ecological Disaster," in *Easter Island Studies: Contributions to the History of Rapanui in Memory of William T. Mulloy*, ed. S. R. Fischer (Oxford: Oxford Books, 1993), 27–45; Catherine Orliac, "The Woody Vegetation of Easter Island Between the Early 14th and the Mid-17th Centuries AD," in *Easter Island Archaeology: Research on Early Rapanui Culture*, ed. C. S. Ayres and W. Ayres (Los Osos, CA: Easter Island Foundation, 2000), 211–20; and Valentí Rull, "The Deforestation of Easter Island," *Biological Reviews* 95 (2020): 124–41.

5. See John E. Randall and Alfredo Cea, *Shore Fishes of Easter Island* (Honolulu: University of Hawaiʻi Press, 2011) regarding shore fishes, and the ethnological accounts of Katherine Routledge, *The Mystery of Easter Island* (London: Hazel, Watson and Viney, 1919); Alfred Métraux, *Ethnology of Easter Island* (Honolulu:

Bishop Museum Press, 1971); and Sebastián Englert, *La tierra de Hotu Matu'a: Historia y Etnología de la Isla de Pascua*, 9th ed. (Santiago: Editorial Universitaria, 2004, 1948) for discussions of prehistoric offshore fishing.

6. Earlier linguistic and archaeological approaches to dating the arrival of settlers yield estimates of 400 to 800 CE. See William Mulloy and Gonzalo Figueroa, *The A Kivi-Vai Teka Complex and its Relationship to Easter Island Architectural Prehistory* (Honolulu: Social Science Research Institute, University of Hawai'i at Mānoa, 1978). More recent investigations take the various fields of paleoecology into account. See Jo Anne Van Tilburg, *Easter Island: Archaeology, Ecology and Culture* (London: British Museum Press, 1994); John R. Flenley and Paul Bahn, *The Enigmas of Easter Island* (New York: Oxford University Press, 2002); John R. Flenley and Paul Bahn, "Conflicting Views of Easter Island" *Rapa Nui Journal* 21, no. 1 (2007): 11–13; Daniel Mann et al., "Prehistoric Destruction of the Primeval Soils and Vegetation of Rapa Nui (Isla de Pascua, Easter Island)," in *Easter Island: Scientific Exploration into the World's Environmental Problems in Microcosm*, ed. John Loret and John T. Tanacredi (New York: Kluwer Academic–Plenum, 2003), 133–53; Daniel Mann et al., "Drought, Vegetation Change, and Human History on Rapa Nui (Isla de Pascua, Easter Island)," *Quaternary Research* 69 (2008): 16–28; Catherine Orliac and Michel Orliac, *Easter Island: The Mystery of the Stone Giants* (New York: Abrams, 1995); Catherine Orliac, "The Woody Vegetation"; Catherine Orliac, "Ligneux et palmiers de l'île de Pâques du XIéme a XVIIéme siécle de notre ére" in *Archéologie en Océanie insulaire: Peuplement, sociétés et paysages* (Paris: Editions Artcom, 2003): 184–99. Terry Hunt and Carl Lipo, "The Archaeology of Rapa Nui (Easter Island)," in *The Oxford Handbook of Prehistoric Oceania*, ed. Ethan E. Cochrane and Terry L. Hunt (New York: Oxford University Press, 2018): 416–49, assign a significantly later date than most, locating settlement in the mid-13th century.

7. This is William Mulloy's conviction. See his "Contemplate the Navel of the World," *Americas* 26, no. 4 (1974): 25–33.

8. Ben Finney discusses this question in "Tracking Polynesian Seafarers," *Science* 317, no. 5846 (September 28, 2007): 1873–74.

9. See K. R. Howe, *Vaka Moana: Voyages of the Ancestors: The Discovery and Settlement of the Pacific* (Auckland: David Bateman, 2006), 92–98.

10. See Ben Finney, "Voyaging Canoes and the Settlement of Polynesia," *Science* 196 (1977): 1277–85; Geoffrey Irwin, *The Prehistory Exploration and Colonization of the Pacific* (Cambridge: Cambridge University Press, 1992), 216; and Peter S. Bellwood, *The Polynesians: Prehistory of an Island People* (London: Thames and Hudson, 1978), 39.

11. Because of Hawaiian prominence in voyaging studies and demonstrations, use of their term is especially apposite to this discussion.

12. Milton's expression in Sonnet XI could hardly be more apropos. In Book XI, 2:20–25 of the *Institutio Oratoria*, Marcus Fabius Quintilianus (35–100 CE)

discusses the matter of cultivating an extraordinary reservoir of memory precisely by means of spatial referents accessed *visually*.

13. Especially *Gygis alba* and *Anous stolidus*—in Hawai'i, *manu-o-Kū* and *noio koha* respectively; *Pluvialis fulva*, called *tōrea* in Tahiti; *Urodynamis taitensis*, among the Māori, *koekoeā*, or *kārevareva* elsewhere; *Numenius tahitiensis*, the *te'ue*; and, on Rapa Nui, the *Onychoprion fuscatus* and *Onychoprion lunatus*, both of which are known to us generically as *manutara*. See Andrew Crowe, *Pathway of the Birds: The Voyaging Achievements of Māori and Their Polynesian Ancestors* (Honolulu: University of Hawai'i Press, 2018), chapter 5, "Traditional Wayfinding at Sea."

14. Joseph H. Genz, "Resolving Ambivalence in Marshallese Navigation: Relearning, Reinterpreting, and Reviving the "Stick Chart" Wave Models," *Structure and Dynamics* 9, no. 1 (2016): 8–40.

15. Shall we conceive of these segments as *quadrants* according to the better-known concepts of European navigational practice? In her review of Pacific ethnoastronomical data, Maud Worcester Makemson notes that, in Kiribati, "the sky or roof of voyaging, *uma ni borau*, was bisected by the ridgepole (meridian), *te taubuki*, and supported by imaginary rafters, *oka*, three on the east and three on the west, vertical to the horizon. The northern pair met where the Pleiades cross the meridian, 24' north of the celestial equator; the southern pair had their apex where Antares transits, or 26' south of the equator" ("Hawaiian Astronomical Concepts," *American Anthropologist* 40 (July–September 1938): 370–83). As Dennis Kawaharada put it to me in private correspondence, ancient Polynesian navigators seem rather to have conceived of the night sky "as the ceiling of a house with pillars or posts and rafters."

16. David Lewis, *We, the Navigators: The Ancient Art of Landfinding in the Pacific*, 2nd ed. (Honolulu: University of Hawai'i Press, 1994) points out that "steering by horizon stars is every bit as accurate as by magnetic compass and probably easier than trying to follow the gyrating compass card of an island schooner or a yacht. The snag is that the navigator who is going to use the stars as we would a compass must be so thoroughly familiar with the night sky that he can orient himself when no more than one or two stars are visible . . ." (121).

17. Chicken bones initially thought to have carried Polynesian mtDNA haplotypes were found at El Arenal-1, a pre-Columbian site in the Biobío Region of Chile. See Finney, "Tracking Polynesian Seafarers," 1874; Terry Jones et al., *Polynesians in America: Pre-Columbian Contacts with the New World* (Lanham, MD: Altamire Press, 2011), 267; and Alice Storey et al., "Radiocarbon and DNA Evidence for a Pre-Columbian Introduction of Polynesian Chickens to Chile" *Proceedings of the National Academy of Sciences* 104, no. 25 (19 June 2007): 10338–39. Later research, however, challenges these findings. See Vicki Thomson et al., "Using Ancient DNA to Study the Origins and Dispersal of Ancestral Polynesian Chickens Across the Pacific," *Proceedings of the National Academy of Sciences* 111, no. 13 (1 April 2014): 4826–31.

18. For details of the 1999 voyage of the Hōkūle'a from Mangareva to Rapa Nui, see http://archive.hokulea.com/rapanuiback.html, accessed October 29, 2020.

19. Though still compatible with the view that *kumara* came to Rapa Nui after it had been introduced to the central Polynesian archipelagoes, current research now casts considerable doubt upon the hypothesis of a Rapa Nui route toward pre-historic Polynesian landfalls in South America. See Jon G. Hather, "The Archaeobotany of Subsistence in the Pacific," *World Archaeology* 24, no. 1 (June 1992): 75.

20. Pablo Muñoz-Rodríguez et al., "Reconciling Conflicting Phylogenies in the Origin of Sweet Potato and Dispersal to Polynesia," *Current Biology* 28 (2018): 1246–56, "provide evidence that the sweet potato was present in Polynesia in pre-human times."

21. Notwithstanding Robert Langdon, "Manioc, a Long-Concealed Key to the Enigma of Easter Island," *The Geographical Journal* 154, no. 3 (1988): 324–36, recent genetic studies, together with more current research impacting the dates of Rapa Nui settlement, conclude that the transmission of Amerindian mtDNA into Polynesian populations occurred before the settlement of Rapa Nui. See A. G. Ioannidis et al., "Native American Gene Flow into Polynesia Predating Easter Island Settlement," *Nature* 583 (2020): 572–77.

22. See Erik Thorsby et al., "Further Evidence of an Amerindian Contribution to the Polynesian Gene Pool on Easter Island," *Tissue Antigens* 73, no. 6 (2009): 582–85; Erik Thorsby et al., "The Polynesian Gene Pool: an Early Contribution by Amerindians to Easter Island," *Philosophical Transactions: Biological Sciences* 367, no. 1590 (March 19, 2012): 812–19; see also Erika Hagelberg, "Genetic Affinities of the Rapanui," in *Skeletal Biology of the Ancient Rapa Nui*, ed. G. W. Gill and V. Stefan (Cambridge: Cambridge University Press, 2016), 182–201.

23. Though Heyerdahl meant for the 1947 *Kon-Tiki* expedition to demonstrate the plausibility of his theory of diffusionism, Holton (2010) offers a spirited summary of why it didn't. See Thor Heyerdahl, *The Kon-Tiki Expedition* (London: Allen & Unwin, 1950) and *American Indians in the Pacific* (London: Allen & Unwin, 1952); and Graham E. L. Holton, "Heyerdahl's Kon Tiki Theory and the Denial of the Indigenous Past," *Anthropological Forum* 14, no. 2 (2004): 163–81.

24. Jones et al., *Polynesians in America*, 263.

25. See Ben Finney, "The Impact of Late Holocene Climate Change on Polynesia," *Rapa Nui Journal* 8, no. 1 (1994): 13–14.

26. See Lewis, *We, the Navigators*, 298–301.

27. See David Porter, *Journal of a Cruise Made to the Pacific Ocean*, 2nd ed., 3 vols. (New York: Wiley & Halsted, 1822).

28. See Lewis, *We, the Navigators*, 307; and Patrick Kirch, *A Shark Going Inland is My Chief: The Island Civilization of Ancient Hawai'i* (Berkeley: University of California Press, 2012).

29. The islands of Hiva 'Oa, Fatu Hiva, and Nuku Hiva all belong to the group.

30. In the Rapa Nui language, *kuhane*. Elsewhere in Polynesia, spiritual beings often present themselves as pelagic fauna, or *nā 'aumākua*, as they are called in Hawai'i.

31. See Kirch, *Shark Going Inland*, chapter 3, for a typical example.

32. For other reasons, linguist and ethnologist Steven Roger Fischer comes to the same conclusion. See "Rapanui's Tu'u ko Iho versus Mangareva's 'Atu Motua: Evidence for Multiple Reanalysis and Replacement in Rapanui Settlement Traditions, Easter Island," *The Journal of Pacific History* 29, no. 1 (1994): 3–18.

33. See William Mulloy, "Double Canoes on Easter Island?," *Archaeology and Physical Anthropology in Oceania* 10, no. 3 (Oct 1975): 182.

34. *'Uru* has since come to Rapa Nui from Tahiti.

35. See Kirch, *Shark Going Inland*, chapter 3 for details of building voyaging canoes in the Marquesas.

36. Voyaging canoes could extend from 15 m to 22 m in length and average 100 to 150 nautical miles in a day (cf. Irwin, *Prehistory Exploration*, 43–44).

37. Sébastien Larrue et al., "Anthropogenic Vegetation Contributions to Polynesia's Social Heritage," *Economic Botany* 64, no. 4 (15 December 2010): 331b.

38. See Patrick Kirch, *The Evolution of Polynesian Chiefdoms* (Cambridge: Cambridge University Press, 1984).

39. Thomas Barthel, for example, provides a relatively spare inventory, but even some of the species he finds attested in the ethnographical record may well have been introduced or reintroduced after European contact; see *The Eighth Land: The Polynesian Settlement of Easter Island* (Honolulu: University Press of Hawai'i, 1978), 129–31. Contrast Barthel's list with Paul Alan Cox and Sandra Anne Banack, eds., *Islands, Plants and Polynesians: An Introduction to Polynesian Ethnobotany* (Portland, OR: Dioscorides Press, 1991).

40. Terry Hunt, "Rethinking Easter Island's Ecological Catastrophe," *Journal of Archaeological Science* 34 (2007): 486.

41. Carl Skottsberg, *The Natural History of Juan Fernandez and Easter Island* (Uppsala: Almqvist & Wiksells Boktryckeri, 1956), 426.

42. The toromiro has since been reintroduced to the island. The plants, grown in Kew Gardens, came from seeds harvested by Thor Heyerdahl in 1955.

43. Formerly designated *Jubaea paschalis* or *Jubaea disperta*, taxonomist and palm specialist John Dransfield reclassified the extinct Easter Island palm as *Paschalococos disperta* on the basis of the hard cocoid endocarp and three eyes found on nuts in caves. Though similar, he specifies that it should not be conflated with the *Jubaea chilensis*. See John Dransfield, John Flenley, Sarah M. King, D. D. Harkness, and Sergio Rapu, "A Recently Extinct Palm from Easter Island." *Nature* 312 (1984): 750–52; and Claire Delhon and Catherine Orliac, "The Vanished Palm Trees of Easter Island: New Radiocarbon and Phytolith Data," in *VII International Conference on Easter Island and the Pacific Aug 2007* (Visby, Sweden: Gotland University Press, 2010).

44. Nor, indeed, would the soft, interior fibers of any kind of palm tree of the *Aracaceae* family be suitable for the hull of ocean-going canoes, though coco fibers were used for sails and rigging. See the preface to Kirch, *Shark Going Inland*.

45. See Delhon and Orliac, "Vanished Palm Trees," 97; and Valentí Rull, "Deforestation of Easter Island."

46. See Patrick C. McCoy, "Easter Island," in *The Prehistory of Polynesia*, ed. J. D. Jennings (Cambridge, MA: Harvard University Press, 1979), for an example of the first tendency, and, for the second, Dennis Kawaharada, "The Discovery and Settlement of Polynesia," Hawaiian Voyaging Traditions, accessed October 17, 2021, http://archive.hokulea.com/ike/moolelo/discovery_and_settlement.html.

47. Van Tilburg, *Easter Island*, reports 887 moai, but Hunt and Lipo, "Archaeology of Rapa Nui," calculate the number at 961 (cf. 419 and 424–25). See also Britton Shepardson, *Explaining Spatial and Temporal Patterns of Energy Investment in the Prehistory Statuary of Rapa Nui (Easter Island)*, PhD diss., University of Hawaiʻi-Mānoa, 2006.

48. The paucity of islands our size compels this parochial comparison. For the record, Nantucket, port of both the historical *Essex* and the fictional *Pequod*, each destroyed by a sperm whale in the Pacific, is 272.6 km^2.

49. Under the command of Richard Ashmore Powell (1816–1892), the Royal Navy plundered the statue in 1868 carrying it away on the *HMS Topaze*. See Jo Anne Van Tilburg, *Remote Possibilities: Hoa Hakananaiʻa and HMS Topaze on Rapa Nui*, British Museum Research Papers 158 (London: British Museum Press, 2006).

50. Hélène Martinsson-Wallin, *Ahu: The Ceremonial Stone Structures of Easter Island*. (Uppsala: Societas Archaeologica Upsaliensis, 1994).

51. Hunt and Lipo, "Archaeology of Rapa Nui," 420.

52. See Thor Heyerdahl and Edwin Ferdon, eds. *Archaeology of Easter Island* (Chicago: Rand McNally, 1961).

53. William Mulloy, "A Solstice Oriented Ahu on Easter Island," *Archaeology and Physical Anthropology in Oceania* 10 (April 1975): 2. See also Mulloy and Figueroa, *The A Kivi-Vai Teka Complex*.

54. William Liller, *The Ancient Solar Observatories of Rapanui: The Archaeoastronomy of Easter Island* (Old Bridge, NJ: Cloud Mountain Press, 1993), 31ff.

55. Liller, *Ancient Solar Observatories*, 17, 19.

56. See Mulloy, "Solstice Oriented Ahu," 2–3; William Liller, "The Megalithic Astronomy of Easter Island: Orientations of Ahu and Moai," *Journal for the History of Astronomy* 20, Archaeoastronomy Supplement (1989): s21–48, and *Ancient Solar Observatories*; William Liller and Julio Duarte, "Easter Island's Solar Ranging Device, Ahu Huri A Urenga, and Vicinity," *Archaeoastronomy* 9 (1986): 39–51; Edmundo Edwards and Juan Antonio Belmonte, "Megalithic Astronomy of Easter Island: A Reassessment," *Journal of the History of Astronomy* 35 (2004): 421–33; and Edmundo Edwards and Alexandra Edwards, *When the Universe Was*

an Island: Exploring the Cultural and Spiritual Cosmos of Ancient Rapa Nui (Easter Island: Hangaroa Press, 2013).

57. Edwards and Belmonte, "Megalithic Astronomy," 424–26.

58. Edwards and Belmonte, "Megalithic Astronomy," 427.

59. Englert, *La Tierra*, 1948.

60. See Charles Love, "The Easter Island Moai Roads: An Excavation Project to Investigate the Roads Along Which the Easter Islanders Moved Their Gigantic Ancestral Statues," Report (Rock Springs: Western Wyoming Community College, 2001).

61. See Terry Hunt and Carl Lipo, *The Statues That Walked: Unraveling the Mystery of Easter Island*, ed. Ethan E. Cochrane and Terry L. Hunt (New York: Free Press–Simon and Schuster, 2011), chapter 5 for a summary of each of these methods excluding the Russians, for whom see Aleksandr Vitalyevich Pestun and Rustem Chingizovich Valeyev, "Шествие каменных голиафов: гипотеза [Procession of Stone Goliaths: A Hypothesis]," in *На суше и на море* [*On Land and Sea*], ed. Boris Borobyov (Moscow: Mysl Publishing, 1988), 412–30; and Pestun, *Моаи острова Пасхи: Инженерные решения древних загадок* [*The Moai of Easter Island: Engineering Solutions to Ancient Mysteries*] (Saint Petersburg: Petersburg XXI Century, 2016).

62. Pavel, Pavel, "Reconstruction of the Transport of Moai," *State and Perspectives of Scientific Research in Easter Island Culture* (1990): 141–44.

63. Pestun and Valeyev, "Шествие каменных голиафов," 412–30.

64. Hunt and Lipo, *Statues That Walked*, chapter 5.

65. Studies published in 2010 found that human habitation generally improved the nutrient levels of the soil. See Thegn N. Ladefoged et al., "Soil Nutrient Analysis of Rapa Nui Gardening," *Archaeology in Oceania* 45 (2010): 80–85.

66. Mulloy and Figueroa, *The A Kivi-Vai Teka Complex*, 1978.

67. Though he did not himself set foot upon the island, Roggeveen's log specifies 134. Nonetheless, Karl Friedrich Behrens (1701-1750), an officer in the landing party, puts the figure at 150.

68. See Paul Bahn and John Flenley, *Easter Island, Earth Island*, 4th ed. (Lanham: Rowman & Littlefield, 2017, 1992), chapter 11, "The Island That Self-Destructed," 245ff; and Jared Diamond, *Collapse: How Societies Choose to Fail or Succeed*, revised edition (New York: Penguin, 2011, 2005), Chapter 2, "Twilight at Easter," 79ff. Diamond famously cites Rapa Nui as a case of "pure ecological collapse" (20).

69. See Edwards and Edwards, *When the Universe Was an Island*.

70. See Mulloy, "Contemplate the Navel of the World," 25.

71. Paul Ehrlich, *The Population Bomb* (New York: Ballantine Books, 1968).

72. Hunt and Lipo, "Archaeology of Rapa Nui," 439.

73. Christopher M. Stevenson et al., "Variation in Rapa Nui Land Use Indicates Production and Population Peaks Prior to European Contact," *Proceedings of the National Academy of Sciences*, 112, no. 4 (2015): 1029.

Chapter Six

Celtic Geometric Art as a Visual Rhetoric of Science

EVELYN DSOUZA

In 2013, Dr. Wenxin Wang and his team of fellow scientists at the National University of Ireland, Galway announced in *Nature Communications* their development of a breakthrough technique in polymerization.[1] This technique, the vinyl oligomer combination strategy, works by controlling polymer growth in slow motion—producing single chains that link repeatedly, wrapping around themselves to form dense structures. These dendritic (or tree-like) hyperbranched compounds can be synthesized in bulk for industrial and medical applications—the improvement of drug delivery, for example, or the creation of elastics and adhesives. As reported in popular and academic Irish news outlets, the research to develop this method of forming polymer chains "took inspiration from ancient arts": specifically, the Celtic knot.[2]

Celtic knots, though various in style and form, are highly recognizable. While some designs (like the *anam cara* knot) are distinctly modern, found most readily at gift shops and tattoo parlors, the art also has an ancient and intercultural history. The symmetries, spirals, key patterns (which are essentially square spirals), and knots that have adorned the walls, monuments, mirrors, shields, swords, and manuscripts of the past have been catalogued extensively in archaeological literature. While these symbols and motifs are fascinating through the lens of art history, they are additionally intriguing as visual-rhetorical sites of scientific and mathematical invention. Beyond the 2013 case of the knot-inspired polymer breakthrough, references to Celtic knots are included in histories of modern-day knot theory.[3] Knot theory, in turn, informs the view in physics of "the universe as a coat of chain mail," which suggests that "at the most fundamental level, space and time themselves are made up of loops—infinite numbers of loops, all linking and knotting and intertwining."[4]

What could these instances of symbolic, cultural, and intellectual uptake offer for interdisciplinary studies of science and technology? As scholarship in the rhetoric of science slowly expands beyond its traditional focus on Greco-Roman linguistic devices and modes of inquiry, ancient and medieval Celtic rhetoric remains relatively unexamined in this body of literature.[5] To be clear, as I will elaborate within this chapter, Celtic geometric art has a fluid cultural history—and even the term "Celtic" has been troubled in the historical and anthropological record. While interlaced visual patterns are evident throughout medieval Europe, Asia, and Africa, and swirls are not just a motif of Migration Period art of late antiquity (the art of Germanic peoples at around 476–800 CE) but of Indigenous art in the Americas and Pacific Islands, this analysis is centered on their extensive use within the Insular art of Ireland and the British Isles—in all its hybridity and intercultural artistic borrowing.[6]

Ultimately, I argue that Celtic topological practices can have meaning for us as visual metaphors for world-making and community-making. Rhetoricians of science have taken a keen interest in the use of visual representation in meaning-making—for example, the graphical depiction of the littoral zone in a geological memoir[7] or circulations of the famous "hockey stick" graph in climate change discourse[8]—and this essay dually examines Celtic geometric art (nonlinguistic, non-discursive rhetorical figures of knots, spirals, and symmetric patterns) as a powerful visual rhetoric and rhetoric of science. This chapter aims to contribute to global rhetorics of science using the trope of metaphor, showing how an ancient visual practice can help us make and interpret arguments in an emergent age of complexity science by shaping and reshaping our thinking.

Celtic Art as a Rhetorical Practice

In keeping with the goals and principles of this volume, terms from Classical rhetoric need not be imposed on artifacts outside that tradition in order to establish those artifacts' characteristics as rhetorical. It may be useful, however, to draw a connection with the method of *rhetorical sequencing* coined by historian of rhetoric Richard Leo Enos.[9] Historians of rhetoric tend to rely on text-based and literary artifacts—especially those that have already been heavily curated and anthologized by previous generations of scholars. In response, Enos has argued that rhetoricians could push those boundaries by seeking out new primary source materials, including the nonlinguistic and the tangible (e.g., pottery, ceramics, and metalwork), better resembling the research methods of archaeologists: "Our task should be to re-co-create the meaning of what is uttered or written as well as the epistemology that is generating such discourse."[10] This process can certainly apply to the visual—that which is not "uttered" or "written," but "portrayed." The aim of "rhetorical archaeology," which Enos sees as the reconstruction "not only [of] the discourse and the cultural context but also the mentalities that are indigenous to the period," is accomplished by four sequential moves:

1. discovering the social, political, and cultural conditions that give rise to cultural values;

2. reconstructing the rhetorical situation or *kairos* that induces discourse;

3. analyzing the actual discourse; and

4. displaying the work in a manner that reconstructs the dynamic interaction of these layers, "much as we do an exhibit at a museum."[11]

While I have not applied these moves in the exact manner they are prescribed here, I have adapted these considerations in my own rhetorical analysis. In general, what could such a heuristic tell us about spirals, symmetries, and knots *as* rhetoric? Enos's considerations, too, are aligned with the central questions of this present volume: over time, how has the symbolic practice of Celtic knots helped its users conceive of the world, move others, and substantiate communities? In this discussion, I will trace the ancient history of Celtic geometric art in broad brushstrokes and then conduct a closer examination of its adoption in mathematics. I end by positioning the Celtic knot as a visual metaphor that helps us describe and appreciate interrelated and nonlinear associations.

Insular Celtic Art through Time: A Synopsis

Unfortunately, much of what is known about ancient Celts comes from Roman ethnography, and a marauding stereotype has persisted in the popular imagination from those sources.[12] Evidence from the material culture, however, has given scholars a far more robust and nuanced view of continental and insular Celtic peoples. In addition, Celtic art and languages are frequently imagined today in terms of national identity: Ireland, Wales, Scotland, Brittany, Cornwall, the Isle of Mann, and Galicia. However, assumptions of anthropological unity and shared identity among groups historically called "Celtic" have been complicated, and the use of this term in a nationalistic sense is decidedly modern. In the 19th and 20th centuries, the term was invoked for cultural and political purposes by those inspired by Western Europe's ancient past—in other words, the Celtic Revival. So, while "Celtic" persists as a cultural category, it is far from a simplistic or homogenous one.

Classical writers did not use the word *Celtae* to describe inhabitants of Britain or Ireland,[13] but tribes who are today identified as Insular Celts are those who inhabited Great Britain (prior to Roman conquest) and Ireland (which had an independent culture). Insular Celtic artifacts adopted the continental La Tène style that followed the Hallstatt period that spanned the 12th century BCE until about the 6th century BCE, to which the extensive use of symmetries (translational, rotational, and axial) can be traced (see figure 6.1).[14] From the 3rd century BCE onwards, continuing with the Roman conquest of Britain in 43 CE and the introduction of

Figure 6.1. An example of rotational symmetry: a decorative knot with 15 crossings, based on a design in George Bain's *Celtic Art: The Methods of Construction*. Source: Wikimedia Commons, accessed November 27, 2020.

Christian influences at around the 5th century (typified in the famous illuminated manuscript of the *Book of Kells*), a distinctly insular Celtic art tradition developed.[15] Of course, with the ebb and flow of influences and points of cultural contact, styles "accumulate, rather than simply replace each other, so that different styles may be found in the same object."[16] Still, however, patterns that are today considered Celtic (i.e., "swirly" decoration) are generally those that emerged from this time and region, although there are even older examples that predate the Celts—such as the triskeles (also known as triple-spirals, or spirals of life) that adorn megalithic tombs at Newgrange and Knowth beginning around 3,000 BCE (see figure 6.2). Although it was not the origin of "Celtic" identity, Ireland remains a relatively continuous source of knowledge about Celtic art and culture, even after its incorporation and fusion into the medieval Christian monastic movement, because it was not conquered by Rome.

In the introduction to *Rethinking Celtic Art*, archaeologists Chris Gosden and J. D. Hill situate their study of Celtic art within a broader movement in anthropology to "focus not on what objects mean, but on what they do in shaping relationships between people. . . . a positive shift away from an emphasis on meaning, which is in any case hard to know, to a stress of effect."[17] Given that the objects condense and embody many histories throughout Europe and beyond, it is perhaps more worthwhile to "analyse *how* [Celtic] art effectively captured the attention of its viewers and educated them into particular ways of seeing the world" (shifting the approach from extensive typology to phenomenology), rather than attempting to chase and locate inherent, static meanings.[18] This is not

Figure 6.2. A triskele, based on motifs found at megalithic tombs in Newgrange, Ireland. *Source*: Wikimedia Commons, accessed November 27, 2020.

unlike the turn within rhetorical studies toward the *effective* aspects of communication (and a communication act's "circulation, transformation, and consequentiality" over space and time)—the belief, in other words, that "an utterance's meanings do not exist *a priori*; rather an utterance's meanings are the consequences it has in the world."[19] Within this same line of thinking, then, rhetorical questions apply: how do these symbols work in strategic, relational, and material ways to interlink their participants?[20]

The edited collection *Celts: Art and Identity* offers a perspective on "Celtic identity" as manifested in the material objects curated for an international Celtic exhibition. In one chapter, Dr. Fraser Hunter argues that although these were useful objects and not merely ornamental, decoration was a key part of their role in both signaling and impressing connections in an international, more-than-Celtic world and marking the high value of objects. Because these patterns specifically adorned objects of high value, it is plausible that they were meant to signal power and status to an international world—to impress, to influence, and to tell stories[21]—an "embodied aesthetics of power."[22] To apply a materialist perspective, these artful objects may not have simply been representative evidence of a connected world—at least for some—but makers of it. The use of interlacing designs in particular would have signaled not "mere copying" but rather "a society open to such change as it chose to accept and adapt on its own terms."[23, 24]

Characteristically, this style of art is aniconic, meaning it resists direct representation or iconism, and contorts realism: multiple creatures, humans and nonhumans, intertwine in these artistic amalgamations (see figure 6.3). While these deep intricacies might suggest a *horror vacui* on

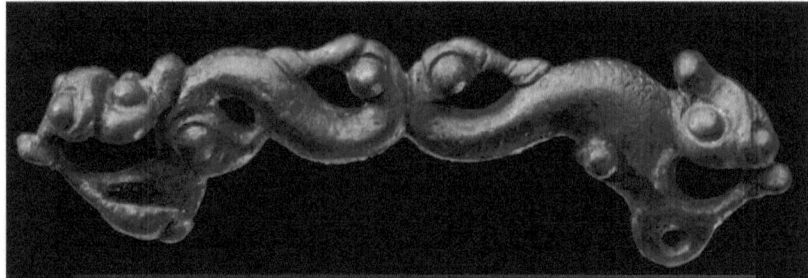

Figure 6.3. A zoomorphic bronze-cast Celtic brooch found in southwestern Germany, about 15 millimeters long. Even in this tiny brooch, ten separate creatures are present—human, bird, and fish intertwined. *Source*: Landesmuseum Württemberg, P. Frankenstein / H. Zwietasch.

the part of the designers, there is an intentionality about them; on even a tiny brooch, for example, metalworkers would have needed a high degree of applied mathematics, as well as artistic and technical excellence. The appearance of these patterns on an object would have signaled something special about that object's display and use.[25]

Celtic Art as a Mathematical Practice

Today, Celtic geometric art endures because of its high degree of cultural transport over time. Its association with history and heritage in the United States and Europe has been popular since the late 20th century—as evinced, for example, by the widespread use of the Celtic cross (or sun cross) in funerary art.[26] While it's clear that these patterns are *artful*, then, in what ways are they also mathematical?

Brent R. Doran offers an in-depth interpretation of Insular Celtic art as a mathematical endeavor and achievement.[27] In several ways, he argues that there is a cultural contrast between Celtic and Greco-Roman forms that betray a difference in worldview; "whereas classical art is rigid, finite, precise, and axiomatic, Celtic art is abstract, iterated, and unbounded—precisely the properties of the more natural and intuitive branches of modern mathematics that have made them such an effective lens to view the complexities of the natural and physical world."[28] By the strategic use of symmetric breaks (i.e., erasing a crossing and rejoining the strand, often in an under-over pattern) and interactive centers of direction and rotational motion in spiral designs, the Celts "have captured a sense of the dynamic . . . reject[ing] the stable, orderly, static designs of their neighbors [in classical antiquity] to find order and pattern amidst the chaos of their spirals," for to them, "the concept of 'center' and locality was a powerful one, dominating much of their thinking."[29] While ambiguity and spellbinding complexity may be the net effects for the viewer, and archaeologists have posited an apotropaic intent for these designs (in that their sheer complexity and multiplicity could act as a snare for demons),[30] Celtic curvilinear art consists of few, simple, self-repeating units that mirror patterns of nature at both micro- and macroscopic levels, like whirlpools and convection currents. This mode of thought and expression contrasts with the Platonic philosophy of harmony that informs Euclidean geometry: Euclidean shapes (lines and planes, circles and spheres, triangles and cones), while "ideal [in] beauty," "turn out to be the wrong kind of abstraction . . . for understanding complexity."[31] In Doran's view, a Celtic paradigm appears "to have embraced the subtly cyclic and occasionally chaotic world around them," which makes it better suited for postmodern scientific thinking[32]—unlike the ancient Greek tendency to demystify nature to intelligence and aim for rationality or realism in art and philosophy.[33]

Laurent Olivier offers a process-oriented interpretation of Celtic visual representation that complements this view. Unlike 19th century scholarship that imagined a hierarchy of artistic representation, wherein stylized or geometric images are deemed "primitive" and the naturalistic work of classical civilizations is deemed more intellectually advanced, Olivier argues that Celtic art is an innovative response to the limitations of both natural vision and optical vision, which "are by themselves always fragmentary and incomplete"—embodying intellectual realism over visual realism.[34] Using both transparency (which renders the essential elements of a figure visible, even if they would not be immediately visible in space) and rabatment (used when a figure requires several views simultaneously to be fully represented in all its constituent elements), Olivier offers a visual reading of Celtic figurative representation that rests "not . . . in the form of occlusions or anatomical rotations—as in naturalistic depictions—but by way of an assemblage of planes, essentially perpendicular . . . This is a form of visual thinking which ignores the distorting effects of perspective, and which allows representation, conditional on the alteration of forms, of the deployment of space of different sides of a single object or figure."[35] In mathematics, a knot is an embedding of a topological circle in three-dimensional space. The triquetra (composed of three interlaced arcs), also known as the trefoil (see figure 6.4), is the simplest mathematical knot and fundamental to knot theory. In Doran's estimation, Celtic artists practiced the only known way to visually (or topologically) depict a three-dimensional knot on a two-dimensional planar diagram without losing information about the object. He writes,

> An aspect of Celtic knotwork immediately striking to the modern mathematically-trained eye is that the knots are all closed, that is, there are no free ends to Celtic knots. Beyond any mystical significance which may have been attached to their closure lies a fairly elementary mathematical truism: to study the nature of a knot, free ends are inessential. Moreover, the "knottiness" is most symmetrically represented in closed form. Essentially, the study of such knots was no more sophisticated

Figure 6.4. A triquetra. *Source*: Wikimedia Commons, accessed November 27, 2020.

than this until the mid-nineteenth century, when the great German mathematician K. F. Gauss addressed their geometry and the Scottish mathematician Tait began his early, tedious attempts at topological classification (up to ambient isotopy).[36]

Fascinatingly, however, an understanding of the mathematics involved "is best gleaned not from the finished product so much as from the methods of construction," namely the introduction of "breaks" in the grid underlying alternating plaits.[37] As documented by J. Romilly Allen, there are eight basic types of knot from which nearly all interlaced patterns in Celtic art emerge: two of these are from a three-cord plait, and the remaining six are from a four-cord plait.[38] However, the evolution of Celtic knotwork out of underlying plaits is not simple, as the underlying plait, which forms knots, can undergo "a bewildering assortment" of modifications and can be easily disguised.[39]

Uptake of Celtic art in present-day texts by mathematicians studying dynamical systems has been largely tacit. As Doran observes, "some authors use [knots] to decorate their texts, although they do not comment further on the designs as art or math."[40] Nonetheless, these ancient patterns continue to yield questions and insights for scientists and mathematicians—even by way of a "deceptively simple question: How many different components (closed loops) are there in a given knotwork design?"[41]

While knot theory is still a nascent area of mathematics that has only made a comeback in recent years, its applications in other disciplines (including the study of DNA, the chirality of molecules, and quantum field theory) have increased its impetus in the field.[42] More broadly, the relationship between topology and ecology—a science that is predicated on dynamic interrelationship—is well-described by ecologists Owen L. Petchey, Peter J. Morin, and Han Olff in the edited collection *Community Ecology: Processes, Models, and Applications*.[43] Here, species are the nodes of ecological networks, and interactions among species are the links. Charting and graphing multiple networks of interactions among species in ecological communities—giving them visuality and shape—help scientists better understand the structural regularities and gaps of these networks.[44]

But why make this connection from topology to ecology—from the "pure" realm of math to messier, material scenes like these? In the concluding section of this chapter, I will reflect on knots as both *symbol* and *substantiation*. Earlier, I emphasized from an anthropological perspective that knots were not just representations of a connected world—they were makers of it. Likewise, from a rhetorical perspective, I see knots as not just as meditative and immaterial representations of real or fantastical things, but as vehicles of power for *actualizing* more flexible and responsive thinking in the face of practical or conceptual problems.

The Celtic Knot as Both Symbol and Substantiation of Ecological Thinking

Complexity, as a framework for scientific investigation and inquiry, emerged in the 21st century as a reaction to the reductionism of Enlightenment-influenced philosophy that emphasized controlled experimentation, typically in a lab setting. In contrast to the linearity and analytical siloing that characterized 20th-century modernism, this approach favors holism and integrates principles of non-linearity, interconnectivity, and interdependence.[45] It accounts for "complex systems and problems that are dynamic, unpredictable, and multidimensional."[46]

While modern rhetorical theory in general will seldom try to reduce communication events or separate them from context, rhetoricians in recent years have taken more interest in complexity itself. An important precursor was Jenny Rice's concept of rhetorical ecology.[47] Building on analyses that had accounted for "a plurality of exigences and complex relations between the audience and a rhetorician's interest," as well as critiques intended to destabilize the supposedly fixed elements of audience, exigence, and constraints, Rice presented rhetorical ecology as an alternative to the more traditional concept of rhetorical situation.[48] "Rather than primarily speaking of rhetoric through the terministic lens of conglomerated elements, I look towards a framework of affective ecologies that recontextualized rhetorics in their temporal, historical, and lived fluxes," she writes.[49] She goes on to offer a networked, rather than fixed, analysis of a public campaign to "Keep Austin Weird" in response to overdevelopment in the city of Austin, Texas; the city is consequently understood as "an amalgam of processes, or as circulation of encounters and actions" rather than a neat container for distinct elements, or a mere backdrop for the events that unfolded.[50] Such a perspective takes into account that "places" are not fixed sites, but points of ever-changing encounter, involving affect, experiences, cultures, histories, moods, and discourses, all embodied—and, importantly, that agency is not located singularly in "the rhetor," but distributed and ecological, in an open rather than closed network. Read this way, rhetorical processes are more like viruses that expand and mutate through new exposures (and Rice specifically invokes here Deleuze and Guattari's description of the "becoming" of evolutionary processes, a sharing and emerging that happens in-between species).[51]

Other rhetoricians have also theorized the relationship between rhetoric and complexity, and in such terms. Chris Mays has written extensively on this subject—recently, for example, in articles for *College English* and *College Composition and Communication* that attend to writing itself as a complex system.[52] Continuing in this vein with a special focus on emerging media and the teaching of writing, Byron Hawk's *A Counter-History of Composition: Toward Methodologies of Complexity* posits a shift in thinking about the romanticist concept of vitalism with analysis rooted in contemporary complexity theory, "transforming rhetoric and composition's image of vitalism from mysticism to complexity" to produce "a post-dialectical understanding of contemporary pedagogies of invention for the emerging science produced by digital technology."[53]

As the basis for invention and inquiry shifts conceptually from reductivism to complexity, and from the linear to the ecological, we need visual rhetorical practices that can adequately convey depth, dimension, density, and interconnectedness—and Celtic knotwork, as geometric, artistic, and community-building practice, may be a useful model toward these ends. After all, decision-making and understanding in all kinds of complex systems rely on modeling, which has "spread far beyond the domains of science and engineering" into such realms as finance and economics, business management, public policy, and urban planning.[54] While much has been written about the persistence of spiral shapes in nature, knots in particular may be useful as conceptual models in the emergent paradigm of complexity science: nodes are discernible, but their beginnings or ends are unclear and cannot be traced to simple or single points of origin.[55]

One moving example of the way that Celtic knots can afford a model for whole-systems thinking—interlinking material and immaterial on the same plane, favoring holism over reductionism—is documented in "The Celtic Knot Project: A Holistic Nursing Intervention Teaching Strategy."[56] Here, nursing educators and registered nurses Drs. Lisa Davis and Sara Bishop describe the problem they encountered: nursing students were having trouble understanding "the holistic nature of nursing," and that "each action by the nurse can affect mind, body, and spirit collectively— that the whole is indeed more than the sum of its parts."[57] They ascribe this problem to the usual formatting of nursing care plans, which requires students to name interventions that are justified by specific and single diagnostic categories. Relying instead on the three-sided Celtic knot to teach "the interconnectedness of the whole. . . . each identifiable but also connected to the other," Davis and Bishop created a new approach for their students to be able to evaluate even simple actions with their patients as an alternative to standard care plans, asking their students to reflect on assessment data, nursing interventions based on those assessments, and a body-mind-spirit rationale for the interventions (using a worksheet or in narrative form).[58] As with the development of the vinyl oligomer combination strategy described earlier in this chapter, an ancient visual resource served not "merely" as decoration or ornamentation—multiplicity for multiplicity's sake—but as an aid and accomplice in the making of an inspired problem-solving approach.

Representations of the knot, as continuous lines with no beginning or end, have appeared persistently in the ways that humans have imagined and communicated concepts like eternity, interconnectedness, and continuity. While such ideas may be as potentially abstract as the graphical illustrations themselves (or more so), it is a systems-level of thinking and imagination that can enhance the way we do rhetorical invention and analysis. Broadly, of course, the Celtic knot and similar figures might serve as companions for the model of rhetorical ecology I have described earlier, referencing the work of Jenny Rice: nodes are interlinked and decentralized, figuratively void of beginnings or ends, but they are discernible and traceable nonetheless. Although Celtic geometric art is a study in "reproducible complexity,"[59] it does what linguistic arguments often cannot do: it expresses what might otherwise be inexpressible, and it works as a sort of gathering space for reflection and meditation, both personal and communal. For rhetoricians of science specifically, the Celtic knot represents one way that searching outside the Greco-Roman canon may generate new rhetorical devices and figures better-suited to communicating complexity. As the epigraph to the introduction of this edited collection states: "We cannot solve our problems with the same level of thinking that created them." In this framework, images of the Celtic knot emerge—or reemerge—as a conceptual model for scientific and technological inquiry and problem-solving, as actualized in the example from Dr. Wang and his team in Galway with which I introduced this chapter. That being noted, as I have alluded to, there is no one "Celtic knot," nor a simple cultural or ethnic origin story. Given the recurrence of these and similar figures globally, this essay should not be interpreted as an argument about Celtic singularity or uniqueness—in fact, the rhetorical connectivity of these patterns works against any such argument. At the same time, any move to decenter the Greco-Roman cultural inheritance of both rhetoric and Euro-American science is a crucial extension of the field(s)—even when that move is made to examine other Western European groups who lived in proximity to and experienced rich and traumatic points of colonial encounter with the Roman Empire.

Conclusion

This chapter has demonstrated how the metaphor of the Celtic knot, as both epistemic representation and ontological substantiation, is a rhetorical resource that can be promisingly reclaimed for the purpose of a more creative and comprehensive approach to modeling technoscientific complexity across communities. Future research will hopefully expand this study's focus on preexisting scholarly commentary and digitized primary source material. My analysis also relied on English-language descriptions of the knots' context; in fact, much of what we know about these patterns remains filtered through a Euro-American academic lens. Important expansions to this inquiry might consider a wider breadth of objects for analysis, including knot-type structures from other Indigenous cultures, and interpret these structures in terms of the worlds and communities they articulated for their designers. This study also invites deeper commentary from scholars with more proximate and indigenous knowledge to the original Celtic creations, who might also share additional or nuancing insights emerging from both English and non-English source material about these powerful and beautiful world-making figures.

Notes

1. Tianyu Zhao et al., "Controlled Multi-Vinyl Monomer Homopolymerization Through Vinyl Oligomer Combination as a Universal Approach to Hyperbranched Architectures," *Nature Communications* 4, no. 1 (May 21, 2013): 1–8, https://doi.org/10.1038/ncomms2887.

2. Jeff Harte, "Ancient Celtic Knots Inspire Scientific Breakthrough," *The Irish Times*, May 21, 2013, https://www.irishtimes.com/news/science/ancient-celtic-knots-inspire-scientific-breakthrough-1.1401644; Anthony King, "Polymer Tied in Celtic Knots," Chemistry World, 2013, https://www.chemistryworld.com/news/polymer-tied-in-celtic-knots/6205.article; National University of Ireland, Galway, "Polymer Breakthrough Inspired by Trees and Ancient Celtic Knots," ScienceDaily, May 22, 2013, https://www.sciencedaily.com/releases/2013/05/130522085335.htm.

3. B. R. Doran, "Mathematical Sophistication of the Insular Celts—Spirals, Symmetries, and Knots as a Window onto Their World View," *Proceedings of the Harvard Celtic Colloquium* 15 (1995): 283; Daniel S. Silver, "Knot Theory's Odd Origins: The Modern Study of Knots Grew out an Attempt by Three 19th-Century Scottish Physicists to Apply Knot Theory to Fundamental Questions about the Universe," *American Scientist* 94, no. 2 (2006): 158–65.

4. M. Mitchell Waldrop, "Viewing the Universe as a Coat of Chain Mail," *Science* 250 (December 14, 1990): 1511.

5. Exceptions include Richard Johnson-Sheehan and Paul Lynch, "Rhetoric of Myth, Magic, and Conversion: A Prolegomena to Ancient Irish Rhetoric," *Rhetoric Review* 26, no. 3 (2007): 233–52, which covers oratory in ancient Ireland (especially the 5th to 9th centuries CE), as well as Paul Lynch, "'*Ego Patricius, peccator rusticissimus*': The Rhetoric of St. Patrick of Ireland," *Rhetoric Review* 27, no. 2 (March 25, 2008): 111–30, https://doi.org/10.1080/07350190801921735.

6. Doran, "Mathematical Sophistication of the Insular Celts," 262.

7. Alan G. Gross, "Toward a Theory of Verbal-Visual Interaction: The Example of Lavoisier," *Rhetoric Society Quarterly* 39, no. 2 (2009): 147–69.

8. Lynda Walsh, "The Visual Rhetoric of Climate Change," *WIREs Climate Change* 6, no. 4 (2015): 361–68, https://doi.org/10.1002/wcc.342.

9. Richard Leo Enos, "The Archaeology of Women in Rhetoric: Rhetorical Sequencing as a Research Method for Historical Scholarship," *Rhetoric Society Quarterly* 32, no. 1 (January 1, 2002): 65–79, https://doi.org/10.1080/02773940209391221.

10. Enos, "The Archaeology of Women in Rhetoric," 70.

11. Enos, "The Archaeology of Women in Rhetoric," 75–76.

12. Fraser Hunter, Martin Goldberg, Julia Farley, and Ian Leins, "In Search of the Celts," in *Celts: Art and Identity*, ed. Julia Farley and Fraser Hunter, 18–35 (London: British Museum Press, National Museums, Scotland, 2015).

13. John Collis, "The Sheffield Origins of Celtic Art," in *Rethinking Celtic Art*, ed. Duncan Garrow, Chris Gosden, and J. D. Hill (Oxford and Philadelphia: Oxbow Books, 2008), 26.

14. Doran, "Mathematical Sophistication of the Insular Celts," 262; Peter R. Cromwell, "Celtic Knotwork: Mathematical Art," *The Mathematical Intelligencer* 15, no. 1 (1993): 36–47.

15. "Celtic Art in Britain, Ireland," *Encyclopedia of Irish and Celtic Art*, accessed December 12, 2022, http://www.visual-arts-cork.com/cultural-history-of-ireland/celtic-art-in-britain-ireland.htm.

16. Chris Gosden and J. D. Hill, "Introduction: Re-Integrating 'Celtic' Art," in *Rethinking Celtic Art*, ed. Duncan Garrow, Chris Gosden, and J. D. Hill (Oxford and Philadelphia: Oxbow Books, 2008), 10.

17. Gosden and Hill, "Introduction," 9.

18. Melanie Giles, "Seeing Red: The Aesthetics of Martial Objects in the British and Irish Iron Age," in *Rethinking Celtic Art*, ed. Duncan Garrow, Chris Gosden, and J. D. Hill (Oxford and Philadelphia: Oxbow Books, 2008), 60.

19. Laurie Gries, "Iconographic Tracking: A Digital Research Method for Visual Rhetoric and Circulation Studies," *Computers and Composition* 30, no. 4 (December 1, 2013): 334, https://doi.org/10.1016/j.compcom.2013.10.006, citing Kevin Porter, *Meaning, Language, and Time: Toward a Consequentialist Philosophy of Discourse* (West Fayette, IN: Parlor Press, 2006).

20. I use the term *participants* here rather than the more standard *users*, as per Kenneth Burke, *Language as Symbolic Action: Essays on Life, Literature, and Method* (Berkeley: University of California Press, 1966). Symbol "using" describes one relationship-a rhetor creates or receives a symbol and "uses" it thus (to achieve some ends). However, I am also trying to capture the ways that symbols "act" on their own, as well, outside what is imagined in the transmission model of simply sending and receiving communication—citing, for example, Bruno Latour, "Visualization and Cognition: Thinking with Eyes and Hands," in *Knowledge and Society: Studies in the Sociology of Culture Past and Present*, ed. Elizabeth Long and Henrika Kuklick, 6:1–40 (JAI Press, 1986).

21. Fraser Hunter, "Powerful Objects: The Uses of Art in the Iron Age." In *Celts: Art and Identity*, edited by Julia Farley and Fraser Hunter, 80–107 (London: British Museum Press, National Museums, Scotland, 2015).

22. Giles, "Seeing Red," 70.

23. Doran, "Mathematical Sophistication of the Insular Celts," 262, citing Ruth and Vincent Megaw, *Celtic Art: From Its Beginnings to the Book of Kells* (London, 1989), 21.

24. As explained in the reference "Celtic Interlace Designs," *Encyclopedia of Irish and Celtic Art*, accessed December 12, 2022, http://www.visual-arts-cork.com/cultural-history-of-ireland/celtic-interlace-designs.htm, there are two theories as to how the broad category of interlace design was introduced to the Insular Celts. One maintains that this is attributed to the Germanic Anglo-Saxon tradition, which prominently features the zoomorphic style also seen in Insular art. The other theory suggests that Celtic interlace may have been derived from the Middle East, as introduced by religious envoys or missionaries. Here, the similarities between Celtic and Levantine interlace are that they are both unending patterns with no beginnings or ends (unlike Germanic or Continental knotwork, which frequently ends with a head or tail), they both favor curvilinear types, and scribes of both places altered the colors of strands that crossed under the others.

25. Hunter, "Powerful Objects."

26. While the Anti-Defamation League has stated that "the overwhelming use of . . . the Celtic Cross [featuring a long vertical axis and knotwork patterns on its surface] is non-extremist," the shadow of nationalistic and ethnocentric appropriation of Celtic symbols by white supremacy groups remains ("Celtic Cross," Anti-Defamation League, https://www.adl.org/education/references/hate-symbols/celtic-cross, accessed December 7, 2022). Such use of this visual rhetoric, apart from its detestability in general, seems counter to the knot's hybrid, syncretic origins.

27. Doran, "Mathematical Sophistication of the Insular Celts."

28. Doran, "Mathematical Sophistication of the Insular Celts," 258.

29. Doran, "Mathematical Sophistication of the Insular Celts," 270.

30. Giles, "Seeing Red," 61.

31. Doran, "Mathematical Sophistication of the Insular Celts," 279, citing James Gleick, *Chaos: Making a New Science* (New York, 1987), 94.

32. Doran, "Mathematical Sophistication of the Insular Celts," 279.

33. Doran, "Mathematical Sophistication of the Insular Celts," 276.

34. Laurent Olivier, "Les codes de représentation visuelle dans l'art celtique ancien [Visual Representation Codes in Early Celtic Art]," in *Celtic Art in Europe: Making Connections*, ed. Christopher Gosden, Sarah Crawford, and Katharina Ulmschneider, trans. S. Crawford (Oxford and Philadelphia: Oxbow Books, 2014): 50.

35. Olivier, "Les codes," 54.

36. Doran, "Mathematical Sophistication of the Insular Celts," 281–82.

37. Doran, "Mathematical Sophistication of the Insular Celts," 274. Methods of knot construction have had to be reconstructed by later artists, mathematicians, and computer scientists, since information by the original artists has been lost.

38. J. Romilly Allen, *Celtic Art in Pagan and Christian Times* (London: Methuen & Co., 1904), 265–66.

39. Cromwell, "Celtic Knotwork," 38.

40. Doran, "Mathematical Sophistication of the Insular Celts," 281.

41. G. Fisher and B. Mellor, "On the Topology of Celtic Knot Designs," Proceedings of the Bridges Conference: Mathematical Connections in Art, Music, and Science, 2004, 37–38.

42. Jessica Connor and Nick Ward, "Celtic Knot Theory" (undergraduate student project, University of Edinburgh, 2012).

43. Owen L. Petchey, Peter J. Morin, and Han Olff, "The Topology of Ecological Interaction Networks: The State of the Art," in *Community Ecology: Processes, Models, and Applications*, ed. Herman Verhoef and Peter J. Morin (New York: Oxford University Press, 2010), 7–22.

44. Petchey, Morin, and Olff, "The Topology of Ecological Interaction Networks," 11.

45. Lawrence Susskind, "Complexity Science and Collaborative Decision Making," *Negotiation Journal* 26, no. 3 (July 1, 2010): 367–70, https://doi.org/10.1111/j.1571-9979.2010.00278.x.

46. Marjorie MacDonald, "Complexity Science in Brief" (University of Victoria, August 2012), 1, https://www.uvic.ca/research/groups/cphfri/projects/currentprojects/complexity (webpage discontinued).

47. Jenny Edbauer, "Unframing Models of Public Distribution: From Rhetorical Situation to Rhetorical Ecologies," *Rhetoric Society Quarterly* 35, no. 4 (2005): 5–24.

48. Edbauer, "Unframing Models," 6–7.

49. Edbauer, "Unframing Models," 9.

50. Edbauer, "Unframing Models," 12.

51. Edbauer, "Unframing Models," 13.

52. Chris Mays, "'You Can't Make This Stuff Up': Complexity, Facts, and Creative Nonfiction," *College English* 80, no. 4 (2018): 319–41; Chris Mays, "Writing Complexity, One Stability at a Time: Teaching Writing as a Complex System," *College Composition and Communication* 68, no. 3 (2017): 559–85.

53. Byron Hawk, *A Counter-History of Composition: Toward Methodologies of Complexity* (Pittsburgh: University of Pittsburgh Press, 2007), 6–7.

54. Muffy Calder et al., "Computational Modelling for Decision-Making: Where, Why, What, Who and How," *Royal Society Open Science* 5, no. 6 (2018): 2, https://doi.org/10.1098/rsos.172096.

55. Consider, for example, the visual-rhetorical use of a Celtic knot as a figure in Syed A. M. Tofail et al., "Additive Manufacturing: Scientific and Technological Challenges, Market Uptake and Opportunities," *Materials Today* 21, no. 1 (January 1, 2018): 22–37, https://doi.org/10.1016/j.mattod.2017.07.001. The image appeared as figure 1 in this original journal article; four nodes are shown to be intertwined to all the others. Here, the use of a knot is effective as a logical diagram for showing relationships.

56. Lisa A. Davis and Sara Bishop, "The Celtic Knot Project: A Holistic Nursing Intervention Teaching Strategy," *International Journal for Human Caring* 7, no. 3 (October 2003): 8–12, https://doi.org/10.20467/1091-5710.7.3.9.
57. Davis and Bishop, "The Celtic Knot Project," 8.
58. Davis and Bishop, "The Celtic Knot Project," 9–10.
59. Doran, "Mathematical Sophistication of the Insular Celts," 282.

Chapter Seven

This Is a Viral Story about Viral Stories

Image and Graphical Power in
COVID Communication in the Navajo Nation

SUNNIE R. CLAHCHISCHILIGI, JULIANNE NEWMARK,
AND JOSEPH BARTOLOTTA

Encountering Viral Stories

Nestled between Victory Life Church and Nizhoni laundromat, in the median that separates US Highway 64, in the far Northwest community of Shiprock in the Navajo Nation, sat a homemade sign that read "Keep Ur Tł'aa' Home" in bold, black letters on plywood (see figure 7.1).

The little sign, not bigger than an average political lawn sign, drew a lot of attention from local and national media, as well as Diné who shuffled in and out of Shiprock. The message garnered both positive and negative responses; some found it was not proper to tell the People to keep their butts home, and others found it necessary. According to an article in a local newspaper, The Navajo Times, the artist behind the sign was a young Diné man, Mattheu Duncan, 19, who was from the community. Duncan said the sign went through a number of revisions and had been removed a number of times thanks to

Figure 7.1. "Keep Ur Tł'aa' Home." *Source:* Photo by Ravonelle Yazzie, *The Navajo Times.*

critics, but he continued to remake the sign and shared it because the message was too important not to share: "Why we wrote that was because, from our point of view, it takes that type of word to catch people's attention to actually stay home."

Since the pandemic landed in the Navajo Nation in early March 2020, many other Diné have found the need to get messages across to community members, visitors, and tourists. They needed people to stay home because the coronavirus had ravaged the Navajo Nation, leaving hundreds dead and thousands infected.

I remember seeing this sign and many others; some pleaded with the People to "Stay Home, Save Lives" and to "Save the Rez" and others reminded them to "Wear A Mask" and to "Protect Our Elders." Some, like Duncan, used the Navajo language, others used cultural images. I remember seeing these signs when I first returned home when one of the first cases of COVID-19 appeared in the reservation. I remember passing the "Keep Ur Tł'aa' Home" sign and letting out a chuckle with my parents while we passed through Shiprock to my mother's family homestead in

Northeast Arizona, in hopes of escaping the virus in the place I called home—Teec Nos Pos, Arizona.

One by one, signs and images popped up all over the 27-square-mile Navajo Nation and in the Twittersphere, and various social media platforms everywhere.

I remember feeling a sense of pride when seeing these homemade signs. It was amazing to see how creative we were and could be even in times of despair. We were resourceful, just as we had always been, and found a way to communicate messages to each other from near and far, and in a way that brought a sense of pride and ownership to the People.

Introduction to Viral Stories and the Science of Health Communication in the Navajo Nation and Beyond

The authors of this chapter anticipate that this work will be quite different in form and content from the other chapters in this volume. This work is multivocal and decenters Euro-American ontologies about scientific and health communication. This work is dialogic, with the consistent voice of our lead author, journalist and rhetoric and writing scholar Sunnie R. Clahchischiligi, reminding readers of the *storying* nature of all of our work. The chapter oscillates between story and critical discourse, using the dynamic forces of communicative methods, and insists that what readers think might be one type of discourse might in fact be another. More to the point, this chapter inhabits a place we might call the "both at once." Our authorial voices blend, bounce, and blur across disciplinary and genre boundaries. The italicized sections represent reflections from our first author and help audiences understand the larger cultural contexts that surround the stories that build this chapter. While this is a chapter in an academic book, it is also a story of research that relies upon and integrates other stories of communities, histories, and relationships. We try to honor the place of each of these elements in this chapter. In this way we are working to tell multiple stories about health science and medicine in Native communities while resisting the academic impulse to reduce those stories to an overarching settler-colonial narrative about the eradication of an invasive disease.

This study considers examples of what the Euro-American tradition has historically called "scientific," or here medical, communication that has emerged within Indigenous communities. The two principal examples in our chapter are COVID-19 visual/textual arguments created by an artist from Zuni Pueblo and a Diné artist who works outside of the Nation. We position these COVID-19 informational communications as themselves conduits of knowledge, behavior, tradition, and cultural ethos. Indeed, the transmission of these informational texts has become viral, in that these have been distributed by individual citizens of Indigenous nations and shared with friends and family across Indian Country, via social media channels, and in material printed form in local contexts. These texts are indeed a continuation of a long tradition of storytelling seen across Native nations. This chapter offers a case study of community-specific and community-generated approaches that appeared in a small sample of Indigenous contexts (Pueblo and Navajo) during the early weeks of the COVID-19 pandemic. Since sovereign Indigenous communities are unique, we cannot generalize the experience described here to all Indigenous communities. Yet, what it reveals is that communities, when faced with crises of community health and well-being, often generate their own health communication strategies, which are not derivative of the otherwise dominant, often governmental, discourses; this is especially relevant in the broader context of this volume. Here, we see Indigenous communities developing practices that do not emulate Western tropes, but where Indigenous narratives *are* the urtext.

When we consider the examples analyzed in this chapter as conveyers of health information, created by and for people who live in communities that have been historically impacted by hegemonic forces dictating practices of so-called "rational" Euro-American medicine, we can understand the need to examine new work of "marginalization" ourselves. This chapter decenters the dominant concepts of "scientific communication" that have embedded within them colonial practices and histories that have strived to reify "rationality," methodologies of individualized medicine, and discourses that separate humans from the social and cultural networks that connect them to the nonhuman forces that have long supported community well-being. We are moving "established" Euro-American discourses and theories of medical and scientific communication to the margins. We are centering Indigenous ways of knowing, sharing, and transmitting information, in this case, critical information about community health amid the global COVID-19 pandemic. Ultimately, we argue that scientific communications that emerge from and disseminate throughout Indigenous communities require greater study. This chapter explores how (some of) these messages operate and the inspiration for their emergence and their creation by Native people who affirmed a crucial and central place as scientific communicators within specific tribes as well as pan-tribally.

Our approach in this chapter follows the guidance of Sisseton Wahpeton Oyate scholar Kim TallBear, who declares that her own approach as a researcher has been to "ponder the politics that run through knowledge production at every stage—how authors and researchers begin where they do, which audiences they imagine will receive their knowledge production, and what leads them to assume that they should research a subject or object."[1] In the case of the communications that we discuss in this chapter, the assumed audiences envisioned by the content creators were Indigenous audiences: sometimes a pan-tribal audience, sometimes a Diné-speaking audience, sometimes an audience that consumes viral online content, such as with images that were intended to spread as far as social media would take them. As TallBear advises, we use this chapter to "ponder the politics" of knowledge production, specifically as related to knowledge (and what we mean by "knowledge" in this context of this book) about care, health, and community in a time of pandemic.

This chapter foregrounds the analytic work of *storying* (an often-used term in Indigenous studies that connotes the active, iterative, always-changing process of story-as-action across place and time) in the context of storying/storytelling traditions about social, health, and political challenges that have faced Diné across thousands of years, and in communications about COVID-19 as hybrid techno-scientific/artistic stories emerged. The authors of this chapter—whose interests span journalism, the teaching of writing, and ethics and usability in technical and professional communication (TPC)—acknowledge that within certain disciplinary silos there have been challenges concerning the integration of non-Euro-American dispositions of communication within research frames and industry or in-the-field applications. This chapter does not foreground Native-community-sited COVID communications as *alternatives* to what Euro-American research might look like. Rather, our work foregrounds these communications in their own (rather than in a comparative) presence, as we explore their own original contexts, applications, and significances. To readers not familiar with Indigenous communication histories and continuing traditions, specifically to those unfamiliar with Diné practices, we offer here a research frame that might read as new to such audiences, as it has been underrepresented in prevailing, mainstream TPC literature. We purposely are not, however, offering a *comparison* between Euro-American scientific COVID communication and COVID communications that have emerged from Diné and Pueblo communities.

The roles of story and narrative within writing studies writ large exemplify the challenges at the heart of our work. In the interest of disambiguation, we use the term *story* to describe, in the words of Rachel C. Jackson (Cherokee Nation of Oklahoma), "rhetorical practices that move across time and space to activate relationships between peoples and places through collaborative meaning making."[2] Such stories tie to other stories and to landscape-as-context. This definition of *story* lies at the core of *storying*, as we use it in this chapter. Indigenous studies scholars and rhetoric scholars often use *narrative* is a less-positive way (though not uniformly). Some use it to describe the "macro-narratives" that have historically governed classrooms, scholarship, and teaching (related to imperialism and colonialism), as Heather Brook Adams has done.[3] Or, as Chadwick Allen has explored in *Blood Narrative*, "master narratives" have tended to subsume the individual, specific experiences of Indigenous people within them, such that diverse experiences (of individual people and individual tribes) are obscured.[4]

Given this context, we feel that what is most germane to our study is work by TPC scholar Angela Haas[5] and American Indian rhetorics/digital rhetorics scholar Kristin Arola (Keweenaw Bay Indian Community Lake Superior Band of Chippewa Indians descendent).[6] Arola's work foregrounding land-based rhetorics and Haas's affirmation that cultural rhetorics scholarship is work "in which subjectivities, positionalities, and commitments to particular knowledge systems are interrelated and situated within networks of power and geopolitical landbases" significantly inform the work that we do in this chapter.[7] Further, we connect our work to the Cultural Rhetorics Theory Lab's assertion that rhetorics rooted in many cultural knowledge-systems are *story*-driven, "fluid and shifting," and necessitate "deliberatively reflexive practice."[8] As these assertions convey, our arguments are necessarily interdisciplinary and move in the ontological direction TallBear encourages.

TallBear acknowledges that scholars in Native American and Indigenous Studies skew toward humanities and literary studies and do not acknowledge the role of "technoscience in Indigenous sovereignty, thus revealing a Eurocentric disciplinary chauvinism."[9] TallBear further argues that Indigenous ontologies do not break the world into disciplines—into "literature," "history," "religion," "biology," "philosophy," "physics"—those categories leading inevitably to hierarchy. Just like breaking the continuum of life into races, sexualities, and species leads to racisms, sexisms, and species-ism. [We] must boldly travel the multiple networks that have arisen in the West after the cutting of the world into knowledge categories. We must agree to be promiscuous disciplinary travelers and radical experimental surgeons, re-attaching knowledges one to another in our approaches to the problems we tackle, hoping their neurological networks will reconnect themselves.[10]

Our work is anchored in Tallbear's ontological project. We believe that hierarchies of medical and scientific communication responding to the COVID-19 pandemic in Indian Country can productively be explored concurrently with a strident commitment to contextualization, historicization, and understanding these image-and-text examples as integrative technoscientific communications.

To use storytelling as a theoretical approach, we invoke Scott Richard Lyons's (Ojibwe/Mdewakanton Dakota) concept of rhetorical sovereignty to present the visual examples in this chapter as story-images, which function in the realm of image-as-medicine; these images are Indigenous medical-communication technologies of the age of COVID-19.[11] The images we share later in this chapter engage community wellness procedures in response to COVID-19 as a means of "owning," in a sovereign sense, multimodal technologies of print-based or hand-drawn content (as in a flyer that can be posted physically, such as on a sign in a median, or in a digital version online) and digital-only communication (as in an image shared on Twitter).

Indigenous Ontologies and Epistemologies

While there is no monolith that is "Indigenous ontology" (as Nations have their own ways of knowing and being in the world), we have chosen to deploy here a pan-tribal ontological frame that is heavily informed by Navajo perspectives. In terms of our chapter's critical and analytic methodology, we have elected to mitigate the presence of non-Indigenous theorists as we try to frame this research. Shawn Wilson, a member of the Opaskwayak Cree Nation, writes about the struggle to compose Indigenous research paradigms within dominant (Euro-American) research paradigms—such as the apparatus, materiality, and distribution model of this very volume.[12] Wilson's acknowledgment of such struggles for scholars is one we hope to amplify.

Our methodology and deliberate preference for Indigenous scholars as our interlocutors for this work indicate the relational paradigm that undergirds this project's exploration of Indigenous scientific communication in the COVID-19 era. As stories themselves are relational (in the senses of being passed down *within* families and, many times, being relatable to those *outside* of a particular family), we espouse a model for organizing research methods expressed by Margaret Kovach, a scholar of Plains Cree and Saulteaux ancestry, who posits that stories themselves are Indigenous research methods. She writes that "story and Indigenous inquiry are grounded within a relationship-based approach to research. The centrality of relationship within Indigenous research frameworks, and the responsibility that evokes, manifests themselves in broad strokes throughout research in the form of protocols and ethical considerations."[13] Our commitment to stories is instantiated from the opening narrative and is continued as narratives are woven throughout the rest of the chapter.

We believe that a study of the stories told in the images that circulated throughout Indigenous communities, which we will discuss soon, can offer another useful valence for better understanding how Indigenous epistemologies and ontologies might offer insights about stronger public health messaging. As Standing Rock Sioux scholar Vine Deloria affirmed in 1970, "We have a chance to build a new cosmopolitan society within the older American society. But it must be done by an affirmation of the component groups that have composed American society."[14] The following analyses are a space for affirming the importance of such storytelling.

Encountering Texts at Home

It was late March when I made my second trip back home to the Navajo Nation. I was driving through Shiprock when I saw graffiti on the front of an old, abandoned building (see figure 7.2). The image was of a somewhat young Diné looking over his left shoulder wearing a respirator and his hair wrapped in a traditional Navajo bun called a tsiiyééł. The text to left of him, written in graffiti, read "BEWARE of COVID-19" and on the right, the direction his head was turned, read "STAY SAFE STAY STRONG EVERYONE T'ahdii kǫ́ǫ́honiidlǫ́" which roughly translates to "We're still here." It was one of the first "homemade" images I came across weeks after COVID-19 came to the Navajo Nation, and it would not be the last. I took a picture of the image, located the artist Ivan Lee, and asked for permission to share it on Twitter. The image was reshared from my page 61 times.

Week after week, Diné artists—professional, amateur, and everyone in between—generated more and more images to get various messages out to Diné as positive coronavirus cases continued to rise.

In preparation for this study, I remembered the dozens of images I came across since March and started compiling them. The collection process began with the images I had shared on my own social media accounts and ones that I had screenshot and sent to family members. I use those images as a base collection, knowing they would lead to other artists and images that made their rounds around Indigenous social media.

My co-authors also compiled graphics they came across in their own circles—Diné and non-Diné—and added them to the file. Next came searches on Google, Twitter, Facebook, Instagram, and Snapchat, using keywords like "Navajo," "Navajo Nation,"

Figure 7.2. COVID-19 graffiti. *Source*: Photo by Sunnie R. Clahchischiligi.

"Navajo Strong," "Indigenous Artist," "Navajo Artist," and the same keywords but with hashtags in from of them. After collecting just over a dozen, we narrowed down the options and discussed each of them, offering questions and interpretations using our various expertise.

The two images selected for analysis in the later part of the chapter warranted many questions and were rich in appearance and message(s). And they were shared on more than one social media platform and generated a strong response from not just the Navajo community but many other Indigenous American communities.

Using an ethnographic approach, I spoke with the artists of our chosen images individually, getting some of their background information before leading into a discussion about their work. Using ethnographic research when conducting research in Indigenous communities is often recommended as it reflects the values and establishes trust with communities.

We chose to reach out to the artists behind the visuals because it is important to us, as it helps make explicit the designers' intentions in representing scientific information within a distinct community.

Context and Virality

We feel it crucial to offer here a succinct summary of the impacts of the COVID-19 pandemic on Diné residing in the Nation to date. The Navajo Nation was hit especially hard by COVID-19. The Nation saw over five hundred deaths due to the virus by the end of September 2020. This land, where latest census figures show 173,667 inhabitants, had over 10,000 confirmed cases in fall of 2020, indicating an infection rate of about 5,900 per 100,000. This infection rate was substantially higher than any US state.[15] However, these data are complicated by the fact that, from the early days of the pandemic, the Navajo Nation was performing more testing per capita than any American state outside New York and Louisiana.[16]

When the virus found its way to the Navajo Nation in early spring 2020, Navajo Nation President Jonathan Nez and the Navajo Department of Health teamed up for the first public announcement to inform the People about the disease. The announcement was made through a town hall meeting broadcast through the local radio station KTNN out of the

Nation's capital in Window Rock, Arizona. Following the announcement, Nez and Navajo Nation officials searched for effective ways to communicate information about the virus, especially its swift spread throughout the country.

Within weeks, the Navajo Nation Department of Health and the established Navajo Nation Health Command Operations Center, a department that would eventually establish itself as a main resource for all things coronavirus in the Navajo Nation, began distributing health and safety information. The command center used many forms of mass communication on the reservation: radio broadcasts, newspapers, brochures, and social media. With social media, this department moved beyond updates through press releases to infographics; some were lists of symptoms that the People could use to determine whether they had the virus and others were reminders to wear a mask, wash hands regularly, and practice social distancing. The infographics were distributed through many different social media platforms and swiftly became a trend. Patrons, tribal and federal government agencies, and local artists eventually joined the efforts to create awareness among the People. Some joined the effort out of personal obligation, others out of community obligation.

While the official response to the pandemic by both tribal and federal leaders was uneven and, in many cases, fell short, members of the Navajo Nation engaged in their own sort of do-it-yourself information campaign to keep up the public consciousness about the seriousness of the pandemic threat. From Facebook to laundromats to roads both physical and digital, members of the Nation started to share visuals they had fashioned, in the form of handwritten signs and computer-generated graphics, which reminded members of their community to stay socially distant, wash their hands often, wear a face covering, and stay at home if they felt sick.[17] To be sure, the Navajo Department of Health undertook its own public information campaign, but in a land as vast as the Nation, even official communications would have a hard time getting the sort of coverage they might in a denser area. The Twitter page for the Navajo Department of Health, which posts useful information a few times per week, was only followed by about 350 other accounts as of mid-October 2020.

Images as Stories and Persuasive Medicine: Our Chosen Texts

The presence of homemade and local public information campaigns, such as those studied below, is an intriguing case of how members of the Navajo

Nation rearticulated scientific information for their own hyper-contextualized situations. To perform this chapter's analysis, we collected a variety of visualizations that were distributed throughout Indigenous digital communities, particularly Navajo-Nation-related realms of social media. Because we are interested in how the designers mediated scientific information in their visuals, we did not think it was important to gather information about how many people actually saw each one. Even though the currency of social media generally concerns the reach of a message, our study is more focused on the way the designers mediated scientific information.

This project's authors began this study by looking at a larger collection of Native American COVID-19 visual communication examples. These visuals came from a cross section of contexts: some were made by tribal or governmental organizations and others were produced by individual artists without affiliations to official organizations. In all cases, the images were circulated on social media (either as born-digital creations or as images of physical communications) to those in Indigenous circles. We decided to focus our collection on visuals that were seemingly designed to be shared on social media. Unlike some of the images shared already in this chapter, showing homemade messages that were placed in public, we expected messages transmitted on social media to speak to larger Indigenous audience by design.

Our chosen two images were not "official" (to deploy a problematic term connoting authority/authorization) COVID-19 communications; rather, they were created by artists in an effort to influence their community without the backing of a bureaucracy. Thus, these images are free of any restrictions or objectives beyond what the artists wanted to compose. The two we selected were chosen for two principal reasons: (1) we were able to contact the original artist for their comments and their permission to reproduce their work; and (2) these images were clearly saturated with tribal references either to the Navajo Nation specifically or to a pan-tribal concept of Indigeneity.

As a part of our data collection, Clahchischiligi reached out to the designers of each selected visual, and, using her training as a journalist, asked for a brief narrative about the artists' thinking in designing their work. This approach was important to us as it helped make explicit the designers' intentions in representing scientific information within a distinct community. We have already written about the importance of storying within Indigenous communities, but there is even further value in the related metadiscourses of these visuals as we attempt to parse how the

Indigenous designers balance their understanding of the audience while trying to communicate scientific information to them.

The chosen examples are immersed in ongoing narratives that navigate aspects of the Indigenous experience in the American Southwest. We are not suggesting that these examples are representative of all Diné or Indigenous experiences, but they should nonetheless give readers a sense of the ontological play that occurs in these Indigenous-created communications that functionally mediate between, and in fact challenge, colonial articulations of science and medicine and the narratives of health and medicine that figure as a common vocabulary in (certain) Indigenous communities.

The Art/Works Collected

Readers should understand that, historically, the expansive size and rurality of the Navajo Nation have made many types of centralized messaging and communication difficult. This reality was exacerbated in a time of pandemic. Most communication on tribal land is made through Navajo radio stations, the tribal newspaper, and social media. However, some of these modes of communication do not penetrate all homes and all families, as many homes still do not have running water or electricity.

Native people often turn to art as a way to respond in times of crisis or despair. Diné, for example, after being forcibly removed from their lands during the Navajo Long Walk in the 1800s, wove rugs that depicted the hurt and suffering felt by the People. When they were eventually allowed to return to their native lands, they spent many decades weaving rugs to tell stories of all they had suffered, including their journey back when many died from hunger, thirst, and violence imposed by American militants.

For some Native people, art is also looked at as a means of healing and cultural expression. Many Native artists, regardless of tribal and national affiliation, share stories of their work and credit it for helping through times of despair. As such, the following examples indicate uses of art as ways to heal and respond.

Visualization 1: Jared Yazzie

Jared Yazzie is the artist behind the visual of the woman sitting in front of a loom with the words "STAY HOME" in turquoise at the bottom of the loom (meaning this is the completed part of the weaving; see figure 7.3).

Figure 7.3. "Stay Home." *Source*: Jared Yazzie.

Yazzie is a Diné graphic artist with a shop that he has run out of Tucson, Arizona, since 2009. His clothing label, OXDX (http://www.oxdxclothing.com), sells screen-printed shirts and art that focus on a Diné aesthetic and speak to Indigenous issues. Yazzie's website states that "OXDX" stands for "overdose," which he feels "describes the state of modern society. Sometimes we need to pull back and remember our culture, tradition, and those who have sacrificed before us." At the bottom of the image of the woman at the loom is the graffiti-stylized script of OXDX.

The graphic of the woman at the loom was not created for the pandemic itself. An earlier iteration of the graphic, called "Grandma Anarchy," was created in 2012 by Yazzie. Initially, as opposed to the weaving work saying "STAY HOME," it had an "A" in a circle, reflecting a popular symbol for anarchy. The "STAY HOME" image, Yazzie said, was first created by hand as a sketch. The medium speaks to his aesthetic as a sketch artist and his desire to create images that can be reproduced, specifically as stencils. Yazzie intends for the images, and the messages behind them, to be reiterated by individual people.

Yazzie said he chose the image in an effort to respond to Diné who were not listening to protocols and pleas from tribal officials and entities. He wanted to remind them of those they were putting at risk by ignoring messages of caution: "Grandma is weaving [a] 'stay at home' rug . . . elders were the ones mostly impacted by the virus. This is a reminder for everyone to not be selfish with the way you're thinking; there's other people's lives at risk." The message asks that Diné stay home, and if they oblige, people like the women in the image will be safe.

As conditions worsened for Diné, a stay-at-home order was put in place in the Navajo Nation, as well as a fifty-seven-hour weekend lockdown. As a result, Jared Yazzie's work as a graphic designer was suspended. Feeling obligated as a Diné artist to help inform other Diné about the severity of the coronavirus and its impact on the People, Yazzie chose to use his skills to become a messenger. He took his 2012 image and altered it to create a message for Diné. The original "Anarchy" sketch, which was a message about Diné women and their independence, Yazzie altered by switching the anarchy symbol into English text that reads "STAY HOME"; he also added a baby-blue surgical mask to the Diné woman's face. The Diné woman in the image represents a Diné grandmother weaving a rug with a message. Though reproduced through computer software, Yazzie's style emphasized the original hand-drawn nature of the sketch.

In Navajo culture, women are looked at as perhaps the most respected in any family, especially elderly Navajo women, because they are matriarchs and Diné are a matrilineal society—bloodline, lineage, and land ownership are all connected to women.[18] Having grown up in the Navajo Nation himself, Yazzie was well aware of the status of women and strategically placed an elderly woman at the forefront of the message. The author's assumed audiences would be able to make the connection between Diné women and the use of them to send an important message.

Diné women are also often associated with weaving; though men also weave, it is women who are depicted in images of Navajos weaving. Placing a Diné woman weaving in his art, Yazzie provides an image that most Diné are familiar with. It also creates a feeling of empathy and comfort because there is a universal understanding that anything concerning Diné women is important and demands attention.

Much of the image is black and white, which speaks to the desire for it to be used as a stencil, but there are two aspects of it that are in color: the mask and the words on the rug. Yazzie said he chose the blue mask because, at the time, it was the most common mask people were wearing, so it placed the image during a specific moment in time. The text, which is also colored baby blue, is the center of the image. The choice to make the words "Stay Home" the same color as the mask created a direct correlation between the two. When audiences view the image at first, their eyes will be drawn to the capitalized, boldface baby-blue text, then to the mask. Audiences see the text first and then the mask, essentially receiving the message to stay home and being told *why* through the image of the mask.

The intended audience is very specific. The image is a Navajo cultural marker. Though weaving is not specific to Diné, the look of the Diné woman adds a layer of specificity. The traditional Navajo *tsiiyééł* bun hairstyle, and the traditional clothing worn by the woman, establish the connection to Diné and culture. The image was reproduced for mass media circulation. The artist chose not to sell the image and created it to inform and caution Diné during a time he thought they needed it. Many Diné social media users reshared the image on various platforms, which suggests that it fulfilled its intended purpose; the message was received.

The image was created in a specific time for a specific audience and was effective at conveying its message, as many shared the image in various media platforms. It also speaks about a specific time in Navajo history and adds to the stories of Diné. It calls attention to a specific issue and asked audience members to act according to the message. Months after the image was released, Diné were able to "flatten the curve"[19] (a graphical metaphor used, in this case, to denote a decrease in the frequency of new case diagnoses of a virus) of the number of new positive cases in the Navajo Nation, and images and messages like this one were credited for helping with the curve, which suggests that Yazzie's image contributed to that period of success.

Visualization 2: Mallery Quetawki

One of the most circulated infographics at the start of the pandemic was created by Mallery Quetawki from the Pueblo of Zuni near Gallup, New Mexico. As an artist-in-resident at the University of New Mexico College of Pharmacy in the community environmental health program, Quetawki noticed that many in her own community were not receptive to messages from tribal and state leaders about the virus. Having lost a family member to the virus, Quetawki turned to her skills as an artist after feeling a sense of tribal and community responsibility. She also turned to art as a way to deal with the loss in her family and other tribal communities. She created an infographic using computer software and distributed it across multiple social media platforms (see figure 7.4).

Figure 7.4. "Be the Social Vaccine." *Source*: Mallery Quetawki.

The above image is rich in Indigenous symbols that represent Pueblos of New Mexico and Diné; other aspects of the image are pan-tribal. Quetawki's graphic is also rich in color, which demands attention. There are also symbols that have come to be associated with the coronavirus.

At first glance, one is drawn to the circles that are lined up on the left, top to bottom. The top circle is the medicine wheel, also known as the Sacred Hoop, which represents spiritual concepts. In this case, it represents the Northern Plains tribes, specifically the Crow and Cheyenne River Sioux in South Dakota. The circle below it is known as the Navajo Ceremonial Basket or the Navajo Wedding Basket. It is used in ceremonies including the Navajo wedding ceremony. The circle at the bottom is a modified Pueblo of Zuni sunflower design. Quetawki said all three images were carefully selected; they represent the communities she and the program work with. The designs in the graphic also attract a larger audience: Indigenous communities, pan-tribally. The designs are tribe-specific, but one or more of them are easily recognizable by others, especially the medicine wheel. The medicine wheel is often adopted by many if not all tribes and modified to represent the values of the tribe or nation that adopts it.

The Navajo basket has added layers of meaning. The design itself is a representation of life. The center of the basket represents the place of emergence. The inner black steps represent clouds and the space between the site of emergence and the clouds represents Earth. The red band represents the sunray and the black steps outside of the red band represent the Holy People from which Diné emerged. The cream outside the border represents the sky. The opening represents Ha'a'aah, East, the dawn that is known as the beginning of life. Holistically, the basket represents Diné's journey through life from birth.

The sunflower Zuni design represents a warrior deity that stands for lineage, a warrior to know the good fights, according to Quetawki. She said the sunflower design also represents strength. To the Pueblo, the sunflower is known for its resilience: though it dries up in the winter, it returns in the spring. To the right of the circle in Quetawki's image, viewers see a modified sunflower against a turquoise background. The sunflower is taken from the circle, singled out, and amplified in the image, representing the Pueblo warrior deity. Embracing the deity is the color turquoise, the color of protection, most associated with the Pueblo and Diné. With the turquoise to protect her, the warrior deity protects Indigenous communities from the coronavirus, which is represented on the right side of the image. As Quetawki described, a line of arrowheads separates the turquoise from the red; the red section represents the virus and the turquoise represents the people. The arrowheads are arrows from the people pointed toward the virus; essentially, the people are using arrows as weapons against the virus. For Diné, arrowheads also represent protection. In this case, the arrowheads are protecting Indigenous communities and their culture(s) from the virus. The arrowheads are also protecting the way of life of Indigenous people everywhere, especially those who are directly represented in the image. About these design choices, Quetawki stated, "It's an overall representation of the idea that we do need to fight this; there are ways to fight."

The final touches Quetawki added to the image were the words. Quetawki strategically used the colors of the medicine wheel in the text. She used words and concepts that are indirectly represented in the graphics, such as "fight" and "protect," which repeat the message that the people need to fight not just for their lives but for their livelihoods, as well as for their cultures. Quetawki does not just offer encouragement to join the fight against COVID-19, but she provides information on how to do so. These instructions function as weapons: avoid crowds, stay home, wash hands, and get tested.

By using signs and symbols that are tribe-specific, Quetawki appeals to tribally specific audiences as well as pan-tribal audiences. Though there was a specific audience in mind, incorporating the medicine wheel expanded the audience to other tribes that would be able to identify it as an Indigenous-specific image. At the time when the image was circulated on social media platforms, the virus had begun to ravage Indian Country, especially the Navajo Nation. Though tribal entities attempted to spread the word about pandemic protocols, Quetawki thought this was not enough. She thought communities she directly worked with needed something with cultural context so that they could understand the severity of the situation: "I wanted to use traditional ways of knowing, to use it visually to reach communities, to translate, to engage communities in need of resources that are already out there, that can be too much for some . . . I feel like I'm indigenizing what's already out there."

As an artist-in-residence, Quetawki typically uses art as a tool to translate science and health information, putting her work into cultural contexts. Quetawki was well aware of the information tribal communities were getting, but also knew that they were not put in a context that allowed for the people to relate to them. Being a member of the Pueblo of Zuni allowed her to use her knowledge and experience to generate an image that would have value. She knew culture and traditional ways of living were threatened by the virus and she wanted others to see that too, so that they might take matters a lot more seriously.

Spreading a Message: Seeing, Hearing, and Feeling #StayHome

As we indicated at this chapter's outset, our intention has been to destabilize—and push to the margin—Euro-American approaches to analyzing scientific communications, particularly concerning virus-mitigation communications in Indian Country. We have done so by working to expand and diversify, rather than reduce and simplify, the stories told about COVID-19 health communication in and around the Navajo Nation. Paraphrasing TallBear, we have traveled promiscuously in this chapter, through disciplines, genre structures, authorial voices, and even "conventions"[20] of terminology. Our readers have traveled with us into images, through stories, and across borders, and we hope our messages have spread, virally, in ways that impact future practices of analysis of scientific communications that are ontologically situated. Ultimately, we hope our readers come to understand the complex interplay between image, text, and design that underlies, and in the end, generates, organic responses to public health crises in local contexts. Our chapter's examples exist within broader considerations of scientific rhetorics on a global scale, which is the theme of this volume. As such, we have shared incidences of Indigenous scientific communication in their own right, with their own narrative dimensions and conceptual integrations resoundingly in the foreground.

> *As Diné braced themselves for another wave of positive cases and COVID-19 related deaths in the late fall of 2020, a new wave of homemade signs, visuals, and memes have emerged. Some have a hint of Indigenous humor and others have a touch of sincerity that only the people in the communities can relate to. While scrolling through one of my social media sites, a friend shared a meme created within the Navajo community, it read "Your belly button is not at Walmart. #StayHome" and underneath the text was a location icon with a heart in the center. Those who are not Diné or might have very little knowledge of Navajo culture might have found themselves perplexed, but not me.*
>
> *Among Diné, one's umbilical cord and its location have a great deal of importance. My late grandmother once told me that when a Navajo baby's umbilical cord falls out it must be buried near the home so that when the child grows and leaves the home they will always return home, where their umbilical cord is buried, to the land where their umbilical cord is tied. The meme is in reference to that and resonates with many Diné, including myself. Since and before the pandemic landed in Navajo land, I've returned home to help my family, where my late grandmother kept my umbilical cord.*

Notes

1. Kim TallBear, "Indigenous Bioscientists Constitute Knowledge across Cultures of Expertise and Tradition: An Indigenous Standpoint Research Project," in *Re:Mindings: Co-Constituting Indigenous, Academic, Artistic Knowledges*, ed.

May-Britt Öhman, Hiroshi Maruyama, Johan Gärdebo (Uppsala, Finland: The Hugo Valentin Centre, 2014), 173–91, https://urn.kb.se/resolve?urn=urn:nbn:se:uu: diva-383415.

2. Rachael C. Jackson and Dorothy W. DeLaune, "Decolonizing Community Writing With Community Listening: Story, Transrhetorical Resistance, and Indigenous Cultural Literacy Activism," *Community Literacy Journal* 13, no. 1 (Fall 2018): 37–54.

3. Heather B. Adams, "Decolonizing Projects: Creating Pluriversal Possibilities/ This Fractured Pedagogy: Listening and Learning from My Best Teacher," *Rhetoric Review* 38, no. 1 (January 2019): 18–22, https://doi.org/10.1080/07350198.2019.1549402.

4. Chadwick Allen, *Blood Narrative Indigenous Identity in American Indian and Maori Literary and Activist Texts* (Durham, NC: Duke University Press, 2002).

5. Angela M. Haas, "Wampum as Hypertext: An American Indian Intellectual Tradition of Multimedia Theory and Practice," *Studies in American Indian Literatures* 19, no. 4 (December 2007): 77–100, https://www.jstor.org/stable/20737390.

6. Kristin L. Arola, "Indigenous Interfaces," in *Social Writing/Social Media: Publics, Presentations, and Pedagogies*, ed. Kristin L. Arola (Fort Collins, CO: WAC Clearinghouse, 2017), 211–26, https://wac.colostate.edu/docs/books/social/chapter11.pdf.

7. Casie Cobos, Gabriela Raquel Ríos, Donnie Johnson Sackey, Jennifer Sano-Franchini & Angela M. Haas. "Interfacing Cultural Rhetorics: A History and a Call," *Rhetoric Review*, 37, no. 2 (2018): 139–54, https://doi.org/10.1080/07350198.2018.1424470.

8. Cultural Rhetorics Theory Lab, "Our Story Begins Here: Constellating Cultural Rhetorics." *Enculturation* 21, no. 1 (October 2015): https://enculturation.net/our-story-begins-here.

9. TallBear, "Indigenous Bioscientists," 186.

10. TallBear, "Indigenous Bioscientists," 186–87.

11. Scott R. Lyons, "Rhetorical Sovereignty: What Do American Indians Want from Writing?" *College Composition and Communication* 51, no. 3 (2000): 447–68.

12. Shawn Wilson, *Research is Ceremony: Indigenous Research Methods* (Halifax: Fernwood Publishing, 2008).

13. Margaret Kovach, *Indigenous Methodologies: Characteristics, Conversations and Contexts* (Toronto: University of Toronto Press, 2009).

14. Vine Deloria, *We Talk, You Listen* (Lincoln, NE: University of Nebraska Press, 1970).

15. Mark Walker, 2020. "Pandemic Highlights Deep-Rooted Problems in Indian Health Service," *New York Times*, October 1, 2020, 8, https://www.nytimes.com/2020/09/29/us/politics/coronavirus-indian-health-service.html?searchResultPosition=1.

16. Zac Podmore, "Navajo Nation Has A Higher Coronavirus Testing Rate Than Utah and Most States," *The Salt Lake Tribune*, April 19, 2020.

17. Pauly Denetclaw, "Young Signmakers: Whatever You Choose to Call It, Keep It Home!" *Navajo Times*, April 9, 2020, https://navajotimes.com/coronavirus-updates/young-signmakers-whatever-you-choose-to-call-it-keep-it-home/.

18. Sunnie R. Clahchischiligi, "Navajo Elders: Alone, Without Food, in Despair." *Searchlight NM*, August 6, 2020, https://searchlightnm.org/navajo-elders-alone-without-food-in-despair/.

19. Siobhan Roberts, "Flattening the Coronavirus Curve: One Chart Explains Why Slowing the Spread of the Infection is Nearly as Important as Stopping It." *The New York Times*, March 27, 2020, https://www.nytimes.com/article/flatten-curve-coronavirus.html/.

20. Tallbear, "Indigenous Bioscientists."

Chapter Eight

A Rhetoric of the Home Ground

Local Knowledge and Data Gathering
among the North Atlantic Glaciers

Ryan Eichberger

Prologue: The Fall of Ice

What I remember most about that morning is the ducks. It was two hours after midnight, and I was perched high on a hill, my back to the Atlantic Ocean. Before me, a glacier glowed with early light, for summer has no night this near the Arctic Circle. Nestled at the glacier's edge was a lagoon where hundreds of icebergs drifted slowly seaward. Some were tiny and crystalline. Others were pale blue crags. Everything was still.

And then something changed, a vibration on the edge of awareness. The ducks—common eiders—exploded upward, winging away toward the Atlantic. The vibration grew into a low rumble, punctuated by pops and whipcracks like the rebound of a metal cable. Across the lagoon, ice at the glacier's edge buckled and fell. A plume of white frost rose skyward, and icebergs bobbled on sloshing waves. In the ensuing silence, an eider cooed.

A shift in my vision happened then, similar to an experience Scottish nature writer Nan Shepherd describes in her book *The Living Mountain*. One day, walking in her native Cairngorm mountains, Shepherd sees a hill close enough to touch. "But it couldn't be," she protests, noting that a loch lay between her and it. "There wasn't room."[1]

So it was for me. The rising frost appeared so close that I felt like one step forward might carry me kilometers across the lagoon. For weeks afterwards, I tried to put words around that vertiginous feeling, expansively bright and absolutely gutting, before accepting that I could not bring it into language. I was missing some conceptual grammar to process the event analytically, let alone emotionally.

My failure of imagination happened at Jökulsárlón, a lagoon fed by the Icelandic glacier Breiðamerkurjökull. My encounter was brief, but Icelanders' relationship with glaciers is long. When the island was settled in the 870s CE by Norwegians fleeing political upheaval, they brought glacier knowledge with them.[2] When the nation established a commonwealth in 930, the first parliament met at Þingvellir, a place of lava rock shaped by bygone glaciers. And shortly after the island swore fealty to the Norwegian crown in 1262, the island's glaciers began to expand.[3] Both local and continental European knowledge of Iceland's glaciers grew in the 18th century, and glaciology flourished from the 19th century through Iceland's independence in 1944 and beyond. Today, the island is home to over 250 glaciers covering 11 percent of the country, yet glaciologists expect nearly all will disappear within two centuries.[4] Iceland looks toward a future without ice.

Iceland's climate story resembles those told throughout the Subarctic and Arctic, the latter of which is warming faster than anywhere in the world. A global increase of 2°C would, for example, spell a 4–7°C increase in the Arctic.[5] Dwindling sea ice is transforming the region's ecologies: in the upper Bering Sea, conditions may soon no longer support species like spectacled eiders and walruses.[6] Hunting will thus become more limited, reducing food security.[7] The amplification of Arctic temperatures is also predicted to increase drought, coastal erosion, and extreme weather in lower latitudes.[8]

Iceland's glaciers have become an icon of these global changes. Photos of retreating ice appear often in public media, many borrowed from glaciological research. Both glaciers and the institutional science through which they are often known have become synecdochic—things standing in for whole systems, as though glaciers were climate change embodied, the clarion voice of Science itself pleading for action. Yet there are other ways to know a glacier. This point was made for me while talking with veteran photojournalist Michael Kienitz not long after I watched the ice fall. I met Michael at a gallery of his glacier images, which he had spent five years capturing in Iceland. Visions of ice gleamed from every wall in blue scoops and whorls. Michael's project had begun one day while sitting in a café in Iceland. A waitress had looked out the window, saying, "From my grandmother's house, they used to see the glacier. Now, you can see almost thirty miles down the coast." Soon after, Michael turned his lens to the retreating ice.

"For the next few generations who will never see glaciers," Michael told me, "these pictures will be a record."

Michael's project captures the sorrow of glacier transformation. As anthropologist Kirsten Hastrup notes, Icelandic landscapes are "part of the local memory and hence of the sense of Icelandicness."[9] Iceland's glaciers exist in a web of microclimates and seabirds, herbaceous flora and human love and grief. They can be known partly by institutional science, but localist practices like hillwalking, mountaineering, farming, and shepherding have generated their own glacier knowledge. Glaciers, and knowledge of them, might best be thought of as metonymic—existing in irreducible entanglement with ecology and, respectively, with other knowledge systems.

And so, reflecting on the fall of ice, I began to wonder: how *does* one come to know a glacier?

On Home Grounds

This chapter offers four stories, which I have assembled through archival work and interviews with environmental researchers familiar with the North Atlantic glaciers, especially Iceland's. Each story focuses on how people deliberately or incidentally gather data and form knowledge through what nature writer Robert MacFarlane calls "long acquaintance with a single place."[10] In other words, each involves someone's *home ground*.

We urgently need ways to deepen our acquaintance with our home grounds. Rethinking data gathering as not merely the domain of academic European or American science but something we do in everyday life—as people did with glaciers long before institutional science took an interest—is one way to deepen that acquaintance. To record changes in one's home ground not only builds knowledge by which to understand and steward local ecologies; it can also help retool outdated environmental mindsets. As environmental historian William Cronon argues, we sometimes forget that "the tree in the garden is in reality no less other, no less worthy of our wonder and respect, than the tree in an ancient forest that has never known an ax or a saw."[11] Cronon encourages us "to honor the wild" at hand, directing wonder and respect for autonomy toward that tree in the garden.[12] His words find a moral companion in botanist Robin Wall Kimmerer's call "to become naturalized to place," to live in gratitude and reciprocity "as if this is the land that feeds you."[13] There is a need to better know our places and to develop, as environmentalist Wendell Berry suggests, a "language of familiarity, reverence, and affection by which things of value are ultimately protected."[14] Learning to notice the everyday mundane has been vital to the growth of glacier knowledge, and is necessary climate work today.

To my eye, MacFarlane, Cronon, Kimmerer, and Berry all circle the idea of *local ecological knowledge*: knowledge gained by sustained personal observation of a place. When passed down intergenerationally and integrated into a community's value system, this knowledge is often called *traditional ecological knowledge*.[15] Like institutionalized academic science, local and traditional ecological knowledge involves systematic observation, producing empirical data with illustrative or predictive power.[16] Yet local and traditional ecological knowledge tends to be qualitative, with emotional or spiritual components specific to people dwelling in a place over time.[17] Let me offer a small example of the kind of localist practices that this chapter primarily focuses on: for decades, my grandfather walked the land around his house, learning the location of wild plants for both utility and pleasure. At eighty years old, he told eight-year-old me where the wild asparagus grew. My grandparents and their house have passed beyond my reach, but I still know—and cherish that I know—where it is possible to forage wild asparagus. Even very minor instances of local ecological knowledge can become both an analytical and emotional grammar of one's home ground.

Local and traditional ecological knowledge is global and diverse. In Iceland, farmers have managed wetlands for centuries using local knowledge.[18] In Australia, Aboriginal people recognized how raptors spread wildfire to flush out prey long before institutional science.[19] In North America, Kimmerer describes how the Three Sisters—corn, beans, and squash—have been planted together for millennia to give each other structure, nutrients, and shelter.[20] Rhetoric researchers have increasingly brought such knowledge into, for example, projects related to industrial tree planting in Canada, the Walleye Wars in Wisconsin, indigenous rights surrounding Bears Ears National Monument, and sustainability and restoration practices in New England.[21]

Local ecological knowledge practices also dovetail with the "big table" ethos of contemporary environmental movements, which invite people to bring their distinct backgrounds, skills, and joys to collective climate work.[22] Joy is important: climate work carries mental health risks ranging from exhaustion to grief.[23] A focus on collectivity also runs counter to decades of propaganda that have equated climate action with flipping light switches and reducing carbon footprints (thereby relieving the corporations and institutions most culpable for climate change of the need to do anything about it).[24] Instead, environmental leaders like marine biologist Ayana Elizabeth Johnson and climate strategist Katharine K. Wilkinson call for action that pairs "scientific rigor and moral clarity, analysis and empathy, strategy and imagination."[25] In other words, different traditions, experiences, and attunements of knowledge must be integrated. Local ecological action does this work.

There is a word for the "big table" ethos shared by local ecological knowledge and contemporary environmental movements: *commensality*, the quality of gathering together at a table to share a meal. Commensality implies people converging on common ground, sharing traditions and imagination rich with gratitude, group resilience, and the memory of ancestors. Commensality nicely describes the sharing often involved in local ecological knowledge. Yet not everyone has access to such experiences; economic disparities and histories of racism limit equal access to places like the one where the wild asparagus grew when I was a child.[26] This chapter cannot fix those issues, but I hope the stories below show why broad access to ecologically commensal experience is necessary and meaningful.

Thus, the four stories in this chapter illustrate the commensal entanglement of different knowledge traditions. But why story? As climate change threatens to destabilize borders and geographies, we need stories more than ever. Generalizable studies are eminently useful, but personal, interior stories capture how we have lived during climate change and help us imagine how we might live differently. In his book *Landmarks*, MacFarlane offers the Sussex dialect word *smeuse*—"the gap in the base of a hedge made by the regular passage of a small animal."[27] Knowing the word, I now see *smeuses* everywhere. The word is a story to make the hitherto invisible visible.

And so, for those studying science and rhetoric, or interested in life during climate change, I share the following stories. In them, it is possible to see localist practices entangling irreducibly, metonymically, with European philosophies of science across lifetimes and centuries. From that entanglement emerge lessons for data gathering as an everyday practice—one offering productive space for both joy and grief in a more-than-human world.

The Glaciologist's Story

The year is 1976, and this is Reykjavík, Iceland, before the tourism boom. Twelve-year-old Snævarr Guðmundsson is walking down the street, unaware that he is about to be surprised. Snævarr believes himself alone in his love for mountains because he has never met anyone who studies them. But as he passes a shop window, he sees an advertisement for a book. It is a book about climbing mountains.

"And I thought, 'No, I am not alone!'" Snævarr laughs decades later, sitting in his office in Náttúrustofa Suðausturlands, the South East Iceland Nature Research Center. As a glaciologist, Snævarr researches historical variations in the island's glaciers using tools one might expect, like a geographic imaging system (GIS) framework, but also artifacts one might not, like old photographs and farmers' journals. In a way, this chapter belongs to Snævarr, whose work illustrates the long, historical entanglement of localist practices and institutional European science which have produced glacier knowledge.

After Snævarr saw the ad in the shop, he began visiting glaciers "hanging in patches" among the mountain peaks. Snævarr describes the mountains as places of peace and solitude, an eternal well where he could "go again and again and drink." And so, long before becoming a glaciologist, Snævarr learned about mountains and glaciers as a localist—by visiting them repeatedly and avidly. When he went to university in his forties, those extensive personal observations became useful.

"Everything that had been taught in the university I had seen. I maybe did not understand it when I was in the mountains, but I had *seen* it," Snævarr tells me. "I even sometimes had questions I could not answer that were then answered in the university. So, all this experience from the mountains was helping me to reorient my sight. When I started working as a glaciologist, this old vision was still there. It just follows."

Snævarr's long acquaintance with glaciers provided an intuitive grammar for understanding them. Becoming a glaciologist supplied a missing lexis but did not render his hillwalking knowledge obsolete. Rather, the "old vision" persisted throughout his education. Today, journeying into the mountains remains a way for Snævarr to replenish himself. His academic knowledge of glaciers and his localist's affinity for mountains are an entangled whole.

As our conversation wanders, Snævarr tells me stories of earlier human contact with Icelandic glaciers. I expand on this history below, and through it we will trace a path back to Snævarr and his glaciological work.

THE PHYSICIAN'S STORY

Let us go back to the beginning, when Iceland's first permanent settlers arrived in the late 10th century. The coastline at the time was covered by birch forests, which mostly disappeared by the 12th century due to felling by settlers and browsing by livestock.[28] Within fifty years of settlement, most land on the island had been claimed and inland trails established to connect north and south while circumnavigating the interior icecaps.[29] Early stories from the time suggest that Icelanders understood and respected glaciers. For instance, the *Landnámabók*, or Book of Settlements, tells how two landholders, wizards Loðmundur and Þrasi, tried to divert a dangerous *jökulhlaup* (glacial outburst flood) toward the other's land, but eventually let the water follow the shortest path to the sea.[30] Glaciers also appear in the Icelandic sagas of the 13th and 14th centuries. In *Egils saga Skallagrímssonar*, warrior-poet Egil names a glacier river "Whitewater" because of its milky color, while *Bárðarsaga Snæfellsáss* tells how the titular Bárðr vanishes into the Snæfellsjökull mountain in western Iceland, becoming a patron god.[31] In the celebrated *Grettis saga Ásmundarsonar*, outlaw Grettir—clever, superhumanly strong, and afraid of the dark—descends the glacier Geitlandsjökull into the green troll-valley Þórisdalur, "enclosed by the glacier on all sides and above."[32]

Such stories coincided with the Little Ice Age, which began in the 13th century. At this time, the island's glaciers began to advance; this process would recur, with some reaching a maximum around 1890. In Iceland's far north, glaciers crawled over pasturelands, ruining farmsteads.[33] The advancing glaciers worsened already difficult conditions, while new laws restricted trade and travel through the highlands.[34] Consequently, most Icelanders came to know only the edges of glaciers. Old grammars of landscape yielded to stories of trolls and outlaws in the unvisited highland interior.

However, several people helped change these perceptions. In the late 1600s, Icelandic physician Þórður Þorkelsson Vídalín wrote extensively about the island's glaciers."³⁵ A half century later, Danish lawyer Niels Horrebow was sent to Iceland by the Danish crown to take stock of the island's people, climate, and natural resources, producing an encyclopedic catalogue in which everything from pumice stones to foxes gets its own chapter.³⁶ After Horrebow left the country, Icelanders Eggert Ólafsson and Bjarni Pálsson continued his work, traveling the island in the 1750s to produce a treatise of their home country organized around geographical regions.³⁷ Whereas Horrebow mentions glaciers only briefly, Ólafsson and Pálsson categorize glaciers, describe their expansion, and hypothesize about their mechanics.³⁸ Icelandic glaciologist Helgi Björnsson credits Ólafsson and Pálsson for having "removed much of the fear and superstition that many Icelanders held for the central highlands by climbing mountains and glaciers not far from human habitation" without supernatural retribution.³⁹

The connection between local knowledge and continental European science would grow when, in 1788, young Icelander Sveinn Pálsson sailed to Copenhagen to learn medicine. There, he also experienced the arts and learned natural philosophy, ultimately seeking a grant from the Natural History Society of Denmark to perform a natural history of Iceland. Over three years, Pálsson would document everything he could about Icelandic environment, making a first ascent of the nation's highest peak, Öræfajökull, walking the old herding paths, and wading through glacier rivers.⁴⁰ Through his observations, he produced *Jöklaritið*, a manuscript documenting glaciers, with unprecedented precision. The text, in Pálsson's neat hand, is arguably the first work of glaciology.

Jöklaritið is ardently localist, using Icelandic literature, common wisdom, and firsthand observation to make inferences about glaciers. For example, the manuscript consistently draws on the *Landnámabók* to link sites Pálsson visited with those known during settlement, as well as to describe changing geographies.⁴¹ References to the sagas appear among passages about glacier mechanics.⁴² Pálsson also draws on the memories of Icelanders like an "old learned gentleman" who tells him about how glaciers form based on—Pálsson surmises—personal observations of soil.⁴³ Pálsson reports that residents of southeastern Iceland know that a low fog rising from a glacier river signals rain.⁴⁴ A sense of the elegiac creeps into the text when he asks the reader to remember things lost: "There is still a fenced-in path or road by which the people of Skjaldbreið drove their livestock to this particular common meadow in order not to trample or ruin the meadows of neighbors," he writes, and beckons readers to remember their names.⁴⁵

And there is particular magic when Pálsson describes his own firsthand observations, speculating that winter cod always arrive first on the western coast because they are drawn by the brightness of Snæfellsjökull glacier.[46] When he describes his ascent of Öræfajökull, he pauses to note the beauty of the mountain buttercup on its slopes, describing new petals of "snowy-white," larger blooms of "saffron-yellow," and older blooms of red.[47] He also draws glaciers in profile, revealing their elevation using darker tones for crevasses and ledges, medium tones for exposed rock and moraine, and lighter tones for snow and ice. In one passage, he runs back and forth across frozen sand to understand how trapped air might produce a sound like gunpowder igniting.[48] And there are moments of revelation: at one point, he writes, "I cannot help introducing an idea at this point, which however crazy it may be, occurred to me while I was observing this eastern part of Breiðamerkurjökull before it started moving."[49] He then hypothesizes about glacier mechanics, anticipating Michael Faraday's idea of regelation a half-century later.[50]

In these ways, Pálsson's scientific insight was made possible by his training as well as literature, local knowledge, and foot-worn experience. However, the Natural History Society of Denmark did not publish his manuscript.[51] Glaciologists Richard S. Williams and Oddur Sigurðsson observe that, if it had been published, "Pálsson would have had a 50-year intellectual lead on his European scientific colleagues in terms of substantiated field observations and descriptions of glaciers."[52] Instead, Pálsson would go on to support his family as a physician, eventually settling at Suður-Vík on Iceland's southern coast. The man who had observed the colors of the mountain buttercup spent decades recording the local weather while farming, fishing, and doctoring. In his last years, he would try again to get his manuscript published without luck. He is buried in the Vík churchyard.[53]

However, Pálsson did pass his unpublished manuscript to Scottish missionary Ebenezer Henderson, who cites Pálsson, as well as his predecessors Eggert Ólafsson and Bjarni Pálsson, in his own book *Iceland; Or the Journal of a Residence in that Island*, which corrects and updates knowledge about Icelandic geography.[54] All these works became sources for other travelers, foreign and native, including the Icelander Þorvaldur Thoroddsen, who traveled uninhabited areas of the island in the second half of the 19th century, further expanding geographical knowledge. Meanwhile, farmers in Iceland increasingly kept weather journals, serving as an "*aide memoire*" to care for animals, harvests, and the land.[55] The entanglement of continental European science and local Icelandic knowledge had become a rhythm, and by the 20th century Icelandic glaciologists were drawing on both traditions in their work. This interplay continues—right to the moment when Snævarr shares an unusual photograph with me in the 21st century.

The Glaciologist's Story Continued

The image in front of me is deceptively simple. Beneath a golden sky, vast cords of glacial ice roll against a mountainside. Yet this is not, in fact, a single photo: two small, black-and-white images have been laid over a larger one. Windows into the past, they reveal a bulkier version of the same glacier, Kotárjökull, as it appeared on August 17, 1890, near the end of the Little Ice Age.[56] Snævarr tells me that Icelanders at the time were fifty years behind continental Europeans in mountain exploration, lacking proper equipment and facing environmental hardship. Glaciers like Kotárjökull were thinly documented in this period, so glaciologists turn to old photos like these, taken by English schoolmaster and mountaineer Frederick W. W. Howell.

When Howell traveled to Iceland in the 1890s, he documented everything from bustling Reykjavík to whale bones beached in Seyðisfjörður.[57] His glacier photos caught the greatest public interest, however. As I worked on this chapter, I spent a month pursuing Howell through newspaper archives, uncovering over forty reviews and a dozen advertisements for lectures he gave using the photos. Let us imagine the scene.

Picture a darkened hall in England or Scotland, where Howell usually lectured. A magic lantern projector has been set up. Photography historian Kelley Wilder sets the mood: "The images cast by projection devices like the camera obscura and magic lantern were like jewels—the intense colours mysteriously heightened by the darkness in which they appeared."[58] Now, imagine Howell speaking, black-and-white images of glaciers flickering through the darkness. These images delighted audiences, with a November 25, 1895, review in the *Birmingham Daily Post* praising his "limelight exhibition of rare and beautiful views."[59] When I talk to Snævarr, I ask about Howell's role as a foreigner studying Iceland.

"Icelanders were very poor at this time. They didn't have any cameras or anything like that," Snævarr says. "We are rather used to other people interfering with our island! I think Icelanders have always welcomed people from abroad. From my perspective, it is obvious that these things would not have been captured without Howell."

Howell's images enable glacier research through a process called rephotography, which involves placing two images together to create a comparison.[60] For the Kotárjökull comparison, Snævarr used a digital elevation model to estimate Howell's location based on the photo's center point (such calculations are accurate within 100–200 meters of the original site). There were complications—a prominent boulder in Howell's photo had eroded—but eventually Snævarr took an image that could be compared to Howell's to determine the volume loss of the glacier. The images appear side by side in an article Snævarr and his colleagues produced about Kotárjökull, which concluded that the glacier has lost thirty percent of its volume since its Little Ice Age maximum.[61] Snævarr also produced a variant image for a research presentation, featuring Howell's images nested atop his own. It is this collage that he shares with me. Together, the images contribute useful data to the glaciological record—but also produce unexpected connection.

"It seems to me that he hung out with me all the time," Snævarr says of Howell, who died in 1901 crossing the glacier river Héraðsvötn, Iceland, near where he is buried.[62] Despite the intervening century, Snævarr has stood almost precisely where Howell stood. This is a unique quality of rephotography, which, as Rebecca Solnit writes, always requires a return to "the site of the photographer's choices," including physical movements, observation points, and processes.[63] As a scientific act, rephotography is commensal: it entangles people across time, drawing together their distinct experiences into a collective whole—even though, in this case, those people never actually met.

At this point, Snævarr pauses. "Haven't you tried this on your own?" he asks.

I have. One bright morning in Iceland, my partner and I awoke in a tent below Skaftafellsjökull, an outlet glacier of Iceland's Vatnajökull icecap. At the Visitor Center, I studied a wall display comparing the contemporary glacier with its appearance in 1925, as photographed by Ólafur Magnússon. In the photo, Magnússon's glacier rolled down from the mountains, ending in a meltwater pool that two people on horseback were fording. I took a photo of the display, and soon we were driving west. As we passed beneath Skaftafellsjökull, I saw Magnússon's angle slide by. Pleading my partner's patience, I slipped out of the car and wandered around with the image open on my outstretched phone until what I saw lined up with the scenery in the photo. I took a dozen shots with my DSLR, and we drove on.

Later, I opened Photoshop and dropped one of my photos onto Magnússon's image. As the peaks aligned, I felt a chill. There was the glacier a century earlier in black and white, and there, through the little window of blue sky that was my photo, I could see a much-diminished glacier. Despite my wariness about the spectacle such comparisons can create, spectacle wasn't what held my attention, but the sensation of having stood where someone else stood a hundred years and a few feet away.[64] The presence was piercing.

Later, I tell Snævarr about the experience.

"Yes, at some point you always have to do that," Snævarr says sagely. "You look at how you can maybe fit a mountain in, and a hill in front of it—you usually cannot reach the total truth anyway. But you are not representing that, necessarily. You are representing that *this* is today's view; *this* is the view one hundred years ago. And that's the story, I think."

Later in the conversation, Snævarr tells me about the Kvísker brothers, who own a farm near Öræfajökull. Around 1870, a woman named Guðrún lived at the farm. One morning, she awoke in a bad mood, reporting that she had dreamed of the Spanish ram, a breed introduced to Iceland to try to improve the sheep stock. She explained to the family, including the Kvískers' grandmother, that this was an ill omen; she had dreamed of the ram before, prior to the Hrútárjökull outlet glacier advancing around the 1830s and causing flooding when it dammed a gorge. A few years after this second dream, Hrútárjökull again advanced. Many years later, the Kvískers' grandmother told glaciologists about Hrútárjökull's twin advances, which she remembered due to Guðrún's unusual dream. In this fashion, glacier knowledge was preserved.

In Snævarr's stories, glaciology happens through intergenerational contact, with the investigator deferring to where a mountaineer stood or taking seriously a dream of portentous ungulates. Information that might be otherwise lost or hidden comes into sight by yielding to knowledge and experiences beyond the typical constraints of institutional scientific practice. In the case of Howell's photos, the result is a diachronic database, which Kimmerer defines as "a record of observations from a single locale over a long period of time."[65] Diachronology is a localist grammar. I am reminded of how Nan Shepherd gestures toward the diachronic in *The Living Mountain*, describing her experience of the Cairngorms: "This process has taken many years, and is not yet complete. Knowing another is endless," she writes. "The thing to be known grows with the knowing."⁶⁶

THE ENVIRONMENTAL HUMANIST'S STORY

Diachronology is the going topic when I sit down with Þorvarður Árnason, director of the University of Iceland's research center in Hornafjörður, to hear how he became friends with a glacier.

"I don't really know why, but one fine day in February, I decided to go take a look at this glacier," says Þorvarður—Þorri for short. Þorri had previously been to the glacier, Hoffellsjökull, but never in winter. Some glaciers slither down mountainsides or hang between peaks, but Hoffellsjökull spreads low between rocky spurs.

"Some days before, the level of the lagoon had dropped, so there were hundreds of beached icebergs. And it was possible to walk all the way up to the glacier, which hadn't been possible before." For the first time, Þorri set foot on the ice. He shot hundreds of photos that day.

"And that's when I understood how different the glaciers are in different seasons," says Þorri. For eight years, Þorri visited Hoffellsjökull to document its changes. The resulting images reveal the glacier's retreat—but also subtler rhythms. Depending on the day, the ice looks fuller, lower, muddier, bluer, whiter. The photos form a record, both illustrative and predictive—but they also document a kinship between Þorri and Hoffellsjökull.

"To some extent," Þorri says, "I would even call this particular glacier my friend because it has given me so much over the years."

While that kinship relates partly to his friendship with a nearby farmer, Þorri tells me that if he does not visit a glacier two or three times a week, he feels "quite miserable." While Þorri's project ended when catastrophic recession left Hoffellsjökull too far away to be photographed, he continues his work through drones. He is also working with aerial photographer Kieran Baxter and glaciologist M Jackson on a film, *After Ice*.

As we talk, Þorri describes ice with a clarity born of a localist's long acquaintance. He describes the fluidity of glaciers, how they float and surge, how water tinkles through the ice flakes above or rumbles deep below. His words remind me of climbing the Icelandic glacier Sólheimajökull and listening to the plinks and chimes of water meandering over the surface. Glaciers *sound* alive.

"I was quite aware about climate change before I came here because I am an environmental humanist working with environmental issues for a long time," Þorri explains. "But the reality of it took a few years to sink in. And this was largely due to seeing it with my own eyes—going repeatedly to the same place."

Þorri understood climate change before visiting Hoffellsjökull, but doing so again and again transformed scientific knowledge into something sensate. He tells me he finds the variation of each visit to Hoffellsjökull—moments of solitude, aesthetic pleasure, even danger—sustaining. His images offer testimony to this growing more-than-human kinship. But for Þorri, his affinity with the glacier is not just about his actions.

"The glaciers have the ability—or maybe we should even say agency—to move human beings very deeply through their aesthetic impact, so I think that this very much adds to their aliveness," he tells me. Later in the conversation, he draws humans and ice closer together.

"The only thing I can compare this with is seeing a big, beautiful animal that's wounded and gradually bleeding out—or the beauty of the glacier is being bled out," says Þorri. "Another way that I have worded it is that I felt that the glacier was in my bloodstream."

Þorri's words make my inexpressible wonder and pain at Jökulsárlón come surging up. By bringing the glacier into the human body, he gives the experience of being with a glacier a visceral closeness, an interiority. As traditions of European science and local knowledge have become entangled, and as the analytical and emotional experiences of glacier observation have likewise converged, so Þorri connects glaciers and humans in inextricable intimacy.

"I mean, that's the thing about living beings, whether they're made of ice or water and carbon," says Þorri. "Eventually we all pass away."

The Arctic Guide's Story

Joy and grief alike feature in the final story, which is about falling in love with—and parting ways with—a home ground. But that is not where the conversation begins when I sit down with Kerstin Langenberger, landscape photographer, hut warden, scientist, and Arctic guide. Raised and currently living in Germany, Kerstin got her degree in environmental science from the Agricultural University of Iceland. To her chagrin, she is most famous for a viral 2015 photograph of an emaciated polar bear she took while serving as an Arctic guide. ("Lousy," she calls it; the photo was taken from far away.) Although we talk about the image, the heart of our conversation is Kerstin's relationship with polar regions, and specifically Iceland—"my second home," she calls it.

"I wanted to go to Africa, but I ended up in Iceland," says Kerstin. Iceland became her gateway to the Arctic. "I slowly started to love the emptiness of the Arctic landscape. There are no trees, and the view goes for kilometers to the next mountain, and the air is so crystal clear, and you can drink the water everywhere, and the animals are not very afraid of you," she explains. "That is really something special that I hadn't experienced before."

Kerstin fell in love with the region and has since spent much of her life there, ranging north into the High Arctic in the summers. She brings her varied backgrounds to bear on how she makes sense of the region.

"I am neither only a photographer nor only a scientist nor only a nature-lover. I'm everything altogether," she says. "I really like that mix. I think it helps."

Kerstin's academic training is reflected in the emphasis she places on vertical knowledge: before visiting a place, she researches it extensively, learning everything she can. But vertical knowledge is also her goal once she sets foot in the place itself: she takes joy in making long acquaintance with a landscape.

"In photography, the longer you spend in one place, the better the results are. And it's the same with everything," Kerstin explains. "And if you then take knowledge into it, you realize, 'Oh, there's a combination.'"

By combining research with prolonged firsthand experience hiking, skiing, sailing, and guiding, Kerstin tries to see "the relation of things." And therein lies the rift between her and her adopted home ground.

"I kind of parted with Iceland a few years ago," Kerstin says. "It just hurts going back every time."

Kerstin is referring to Iceland's tourism boom, which has made solitary experiences with the more-than-human world tougher to find. She worries about the environmental losses—ancient moss crushed by the tourists, waterfalls surrounded by overcrowded boardwalks, a sonic landscape drowned by buzzing drones. Iceland has become "an attraction," she says. It's a painful loss—and pain seems to be a fundamental part of what it means to make a home in the Arctic and Subarctic. At one point, she shares with me a photo she took of meltwater torrenting from a towering glacier into the Barents Sea.

"Some glaciers in Svalbard retreat 400 or 500 meters a year," Kerstin explains. "I was there last summer and it's hard to grasp because the glacier front is 10 kilometers long and one-hundred meters high, but then I feel like—wait a moment! Is that an *island* in front of the glacier?'"

In their ablation and retreat, glaciers reveal their underworlds, which rise up as new and nameless lands—a revelation Kerstin calls "utterly scary."

In the words of M Jackson, "Glacier change is transformation."[67] After years of witnessing winters with dwindling sea ice and summers with too much open water too soon, Kerstin has formed a bittersweet grammar for reading ice. She tells me she can look at a photo of a glacier anywhere in the world and know whether it is stable or retreating. Most are retreating.

"An iceberg in itself is a very sad thing—and most people don't realize this," she says.

Home grounds in circumpolar regions are thick with this paradox of affinity: as Þorri's story also suggests, climate transformation can be simultaneously beautiful and tragic. This is made clear as Kerstin describes Iceland's ice caves.

"It's one of the most fantastic things ever," she says, describing glimmering turquoise caves similar to those I remember from Michael Kienitz's gallery. "How can nature create something so beautiful out of something so sad? Because an ice cave is a hollow left by a melting glacier."

But for all the grief, Kerstin holds onto hope. She continues to search for those places of solitary, intimate contact with the more-than-human world. She describes the wonder of being alone for three weeks with a volcano, or being the only human among five hundred thousand penguins.

"Climate change is happening. But we can still do things," she says. "We have an obligation both to the world and to ourselves to stay positive. You cannot help anyone if you're negative."

When I last hear from Kerstin, she is preparing to return to Iceland. Despite their parting of ways, Kerstin and her second home are still fiercely connected. She hopes to find an ice cave again. It has been a long time.

"I'll ski for three weeks through the interior," Kerstin tells me. "There will be no one there. And so I will experience my Iceland."

Data in a Glacier-Bent World

The stories above show things tangling together that are often thought divided: institutional science practices and localist experience; predictive data and emotional affinity; and the human and the more-than-human. The entanglement is often metonymic—that is, an irreducible intermingling. Were local glacier wisdom wiped from history, a centuries-spanning flow of information would collapse, fundamentally changing contemporary glacier knowledge. From settlement-era literature to 18th-century photographs and portentous dreams, all manner of materials outside the boundaries of European science have been crucial to glacier knowledge.

And when the doors of institutions and academies were shuttered against that knowledge—as in Pálsson's case—glacier wisdom nonetheless lived on in the experience of farmers, or was diverted through other channels in the hands of missionaries who continued to correct and expand glacier knowledge.

These stories also reveal that long acquaintance—methodical or incidental—produces knowledge, an intuitive grammar for recognizing and reading the landscape. Hence Snævarr learned glacier features from walking into the hills long before he attended university, and Kerstin learned to read the health of glaciers based on years spent among them. Similarly, Pálsson's work drew on Icelanders' glacier experience, and Þorri's repeat visits to Hoffellsjökull produced a visible record. Conversely, when forces cut people off from the land—whether the Little Ice Age or Iceland's tourism boom—opportunities for knowledge and affinity can be lost.

There are lessons here for data gathering as an everyday life practice. During the pandemic, I began walking often in my neighborhood, and soon started to notice predictive patterns. For example, a pair of goldfinches could be seen bouncing around some coneflowers down the street at 10:00 a.m. most days. I began entering the various bird species I saw into a global database, making them available to researchers. Similar to Lauren E. Cagle and Denise Tillery's call for technical communicators to advocate "for and with" citizens to address climate change, anyone in and beyond rhetoric can help collect, collate, and share such data to deepen local ecological knowledge and make possible community planning and stewardship.[68] Such data gathering can be done within a neighborhood or a global network, but in both cases can be deeply commensal, involving individual people bringing their experiences together to produce something collective and emotionally rich. Indeed, observing birds became my form of self-care, a reprieve that helped me do other climate work—spending time in local greenspaces improves mental and physical well-being.[69] If we want to build climate action founded on human welfare, security, and autonomy, there is radical and urgent need to expand access to such experiences. These stories testify to the useful emotional space—for both joy and grief—opened up by turning a data-oriented eye on one's home ground.

Additionally, the rhetorical choices in these stories depict a more-than-human world lively with force: glaciers do not just retreat and ablate, but become drinking wells, go off like gunpowder igniting, enter into bloodstreams, and loose islands on the world in moments of terrible revelation. Glaciers are grandly impersonal; they are intimately sensate. They become friends, or they distort the migration paths of winter cod.

Glaciers have a weight of inertia that bends the world around them. For humans, the glacier's inertia bends the world toward both joy and grief. This in-betweenness is characteristic of the climate era: there may now never be a time when we do not live between joy and grief, stability and insecurity, the present moment and the half-glimpsed edge of deep-time transformation.

One task of our in-between era will be to learn to listen to all that is not human, for as beautifully put by rhetorician Natasha Seegert, "everything might be speaking."[70] True ecological commensality will involve not just accounting for the more-than-human forces that shape human storytelling and experience, as in this chapter, but striving to notice the rhetorical forces that carry on without human presence, be they between cod and glacier or goldfinch and coneflower. We must, as rhetorician Joshua Trey Barnett writes, "be prepared to lose some control."[71] In doing so we might find some pleasure in a more fragile and ephemeral rhetoric, a world of stimuli and responses we can only partly know.

Making data gathering an everyday life process might permit us to do one last thing: protect what might otherwise vanish. I was reminded of this when Snævarr told me about the Kvísker brothers. Since the 1930s, the brothers had marked the retreat of a nearby glacier with stone cairns. These days, Snævarr has taken over the measuring. Although many older cairns have disappeared, Snævarr used the farmers' journals to rediscover their locations and recover the glacier's story.

"They are getting fewer and fewer, the people that it is still possible to get such information from," he tells me. Old photographs and journals tell stories that institutional science cannot, while institutional science helps keep artifacts and memories alive. The mutualism between them can be both predictive and resonant.

"These people telling this story didn't have a camera, but they have something to share with us," Snævarr says. "They are just human beings, decent people that are just doing their jobs. I don't want to forget these people."

Epilogue: Eiders Returning

What I remember most about that morning at Jökulsárlón is the ducks. After the ice settled, the eiders began to return to the lagoon. Female eiders are a glossy russet, the males black and cream fading to a caramel glow at the breast. They have an unusual relationship with humans: farmers provide them with space to build their nests, and the ducks leave behind the downy feathers used to warm their eggs. The harvested down is lighter than air and crackles with living warmth.

As I watched from the hill that morning under the midnight sun, the eiders dabbled and dozed, but a few further cracks of the icebergs sent some of them winging away again. As they left, they made a westward arc; as they returned, they made an eastward one—but the same arc each time. There was beauty in the call and response between bird and glacier, a pleasurable humility in noticing a hint of a place's grammar that hitherto had been invisible to me—but only a hint. *The thing to be known grows with the knowing,* as Nan Shepherd wrote.[72]

I sat for a while longer. The eiders continued to trace ever-returning arcs over the pale ice, beneath the soft sky, beyond the yellow moon setting into sighing seas.

Notes

1. Nan Shepherd, *The Living Mountain* (Edinburgh: Canongate, 2019), 42.

2. Helgi Björnsson, *The Glaciers of Iceland: A Historical, Cultural, and Scientific Overview,* trans. Julian Meldon D'Arcy (Durham: Atlantic Press, 2017), 130.

3. Terry G. Lacy, *Ring of Seasons: Iceland—Its Culture & History* (Ann Arbor: University of Michigan Press, 2000), 90-91.

4. Oddur Sigurðsson and Richard S. Williams, Jr., *Geographic Names of Iceland's Glaciers: Historic and Modern* (Reston: US Geological Survey, 2008), 2; Helgi Björnsson and Finnur Pálsson, "Icelandic Glaciers," *Jökull* 58 (2008): 365, 381.

5. Warwick F. Vincent, "Arctic Climate Change: Local Impacts, Global Consequences, and Policy Implications," in *The Palgrave Handbook of Arctic Policy and Politics,* ed. Ken S. Coates and Carin Holroyd (Cham, Switzerland: Palgrave Macmillan, 2020), 508.

6. Jacqueline M. Grebmeier et al., "A Major Ecosystem Shift in the Northern Bering Sea," *Science* 311, no. 5766 (2006): 1461, https://doi.org/10.1126/science.1121365.

7. Vincent, "Arctic Climate Change," 512.

8. Vincent, "Arctic Climate Change," 514–15.

9. Kirsten Hastrup, "Icelandic Topography and The Sense of Identity," in *Nordic Landscapes,* ed. Kenneth Olwig and Michael Jones (Minneapolis: University of Minnesota Press, 2008), 53–76.

10. Robert MacFarlane, *Landmarks* (London: Hamish Hamilton, 2015), 71.

11. William Cronon, "The Trouble with Wilderness, or Getting Back to the Wrong Nature," *Environmental History* 1, no. 1 (1996): 24, http://www.jstor.org/stable/3985059.

12. Cronon, "The Trouble with Wilderness," 25.

13. Robin Wall Kimmerer, *Braiding Sweetgrass: Indigenous Wisdom, Scientific Knowledge, and the Teachings of Plants* (Minneapolis: Milkweed Editions, 2013), 214.

14. Wendell Berry, "Life is a Miracle," *Communio* 27 (Spring 2000): 86.

15. Susan Charnley, A. Paige Fischer, and Eric T. Jones, "Integrating Traditional and Local Ecological Knowledge into Forest Biodiversity Conservation in the Pacific Northwest," *Forest Ecology and Management* 246 (2007): 15; Fikret Berkes, *Sacred Ecology* (New York: Routledge, 2018), 247.

16. Robin Wall Kimmerer, "Weaving Traditional and Ecological Knowledge into Biological Education: A Call to Action," *BioScience* 52, no. 5 (2002): 433, https://doi.org/10.1641/0006-3568(2002)052[0432:WTEKIB]2.0.CO;2.

17. Kimmerer, "Weaving Traditional and Ecological Knowledge," 433.

18. Ragnhildur Sigurðardóttir et al., "Trolls, Water, Time, and Community: Resource Management in the Mývatn District of Northeast Iceland," in *Global Perspectives on Long Term Community Resource Management*, ed. Ludomir R. Lozny and Thomas H. McGovern (Cham, Switzerland: Springer Nature Switzerland, 2019), 77–101.

19. Mark Bonta et al., "Intentional Fire-Spreading by 'Firehawk' Raptors in Northern Australia," *Journal of Ethnobiology* 37, no. 4 (2017): 700, https://doi.org/10.2993/0278-0771-37.4.700.

20. Kimmerer, *Braiding Sweetgrass*, 128.

21. Jennifer Clary-Lemon, *Planting the Anthropocene: Rhetorics of Natureculture* (Louisville, CO: Utah State University, 2019), 9; John Koban, "Walleye Wars and Pedagogical Management: Cooperative Rhetorics of Responsibility in Response to Settler Colonialism," *Rhetoric Society Quarterly* 50, no. 5 (2020): 321–34, https://doi.org/10.1080/02773945.2020.1813324; Laura A. Lindenfeld et al., "Creating a Place for Environmental Communication Research in Sustainability Science," *Environmental Communication* 6, no. 1 (2012): 23–43. https://doi.org/10.1080/17524032.2011.640702; Danielle Endres, "Engaging the Nexus of Environmental Rhetoric and Indigenous Rights," *Spectra* 55, no. 1 (March 2019): 24–25; Caroline Gottschalk Druschke and Kristen C. Hychka, "Manager Perspectives on Communication and Public Engagement in Ecological Restoration Project Success," *Ecology and Society* 20, no. 1 (2015): Article 58.

22. Anand Giridharadas, "Does the climate movement need a makeover?" *The Ink*, August 3, 2021, https://the.ink/p/does-the-climate-movement-need-a; Phoebe Neidl, "Why *All We Can Save* Will Make You Feel Hopeful About the Climate Crisis," *Rolling Stone*, September 21, 2020, https://www.rollingstone.com/culture/culture-news/all-we-can-save-book-climate-ayana-johnson-katharine-wilkinson-1062310; Leah Cardamore Stokes, "A Field Guide for Transformation," in *All*

We Can Save: Truth, Courage, and Solutions for the Climate Crisis, ed. Ayana Elizabeth Johnson and Katharine K. Wilkinson (New York: One World, 2020), 341–42.

23. Ashlee Cunsolo and Neville R. Ellis, "Ecological Grief as a Mental Health Response to Climate Change-Related Loss," *Climate Change* 8, no. 4 (2018): 275, https://doi.org/10.1038/s41558-018-0092-2; Susanne C. Moser, "The Adaptive Mind," in *All We Can Save: Truth, Courage, and Solutions for the Climate Crisis*, ed. Ayana Elizabeth Johnson and Katharine K. Wilkinson (New York: One World, 2020), 270–71.

24. Finis Dunaway, *Seeing Green: The Use and Abuse of American Environmental Images* (Chicago: University of Chicago Press, 2015), 82–83.

25. Ayana Elizabeth Johnson and Katharine K. Wilkinson, eds., *All We Can Save: Truth, Courage, and Solutions for the Climate Crisis* (New York: One World, 2020), xviii–xix.

26. Jenny Rowland-Shea et al., "The Nature Gap: Confronting Racial and Economic Disparities in the Destruction and Protection of Nature in America," *Center for American Progress,* July 21, 2020, https://www.americanprogress.org/issues/green/reports/2020/07/21/487787/the-nature-gap.

27. MacFarlane, *Landmarks,* 5.

28. Kevin P. Smith, "*Landnám*: The Settlement of Iceland in Archaeological and Historical Perspective," *World Archaeology* 26, no. 3 (1995): 320–36, https://doi.org/10.1080/00438243.1995.9980280.

29. Kevin J. Edwards et al., "Landscapes of Contrast in Viking Age Iceland and the Faroe Islands," *Landscapes* 6, no. 2 (2005): 78; Anna Dóra Sæþórsdóttir, C. Michael Hall, and Jarkko Saarinen, "Making Wilderness: Tourism and the History of the Wilderness Idea in Iceland," *Polar Geography* 34, no. 4 (2011): 255, https://doi.org/10.1080/1088937X.2011.643928.

30. Ari Þorgilsson, *Landnámabók* (Reykjavík: Sigurður Kristjánsson, 1891), 179–80.

31. Valdimar Ásmundarson, *Egils saga Skallagrímssonar* (Reykjavík: Kostnaðarmaður, Sigurður Kristjánsson, 1892), 73; Guðbrandur Vigfússon, *Bárðarsaga Snæfellsáss, Viglundarsaga: Þorðarsaga, Draumavitranir, Völsaþáttr* (Copenhagen: det nordiske Literatur-Samfund, 1860), 12.

32. Richard Constant Boer, ed., *Grettis saga Ásmundarsonar* (Magdeburg: Niemeyer, 1900), 221-222.

33. Björnsson, *The Glaciers of Iceland,* 348.

34. Sæþórsdóttir, Hall, and Saarinen, "Making Wilderness," 256.

35. Björnsson, *The Glaciers of Iceland,* 576.

36. Niels Horrebow, *Tilforladelige Efterretninger om Island* (Copenhagen, 1752).

37. Astrid E. J. Ogilvie, "Local Knowledge and Travellers' Tales: A Selection of Climatic Observations in Iceland," *Developments in Quaternary Science* 5 (2005): 277, https://doi.org/10.1016/S1571-0866(05)80013-2.

38. Eggert Ólafsson and Bjarni Pálsson, *Travels in Iceland* (London: Barnard and Sultzer, 1805), 126.
39. Björnsson, *The Glaciers of Iceland*, 149.
40. Richard S. Williams, Jr. and Oddur Sigurðsson, "Editor's Introduction," in Sveinn Pálsson, *Draft of a Physical, Geographical, and Historical Description of Icelandic Ice Mountains on the Basis of a Journey to Them in 1792-1794*, trans. Richards S. Williams, Jr. and Oddur Sigurðsson (Reykjavík: Icelandic Literary Society, 2004), xix-xx.
41. Pálsson, *Draft*, 82.
42. Pálsson, *Draft*, 57, 79, 93-94, 100-3, 129-30.
43. Pálsson, *Draft*, 13.
44. Pálsson, *Draft*, 24.
45. Pálsson, *Draft*, 119.
46. Pálsson, *Draft*, 37.
47. Pálsson, *Draft*, 66.
48. Pálsson, *Draft*, 31.
49. Pálsson, *Draft*, 60.
50. Richard S. Williams, Jr. and Oddur Sigurðsson, "Endnotes," in Pálsson, *Draft*, 162.
51. Williams and Sigurðsson, "Editor's Introduction," xviii.
52. Williams and Sigurðsson, "Editor's Introduction," xxvi.
53. Williams and Sigurðsson, "Editor's Introduction," xxii.
54. Ebenezer Henderson, *Iceland; Or, the Journal of a Residence in That Island, During the Years 1814 and 1815* (Edinburgh: Oliphant, Waugh, and Innes, 1818), 237; Björnsson, *The Glaciers of Iceland*, 158.
55. Ogilvie, "Local Knowledge and Travellers' Tales," 275.
56. Snævarr Guðmundsson, Hrafnhildur Hannesdóttir, and Helgi Björnsson, "Post-Little Ice Age Volume Loss of Kotárjökull Glacier, SE-Iceland, Derived from Historical Photography," *Jökull* 62 (2012): 97.
57. Frederick W. W. Howell, *Icelandic Pictures Drawn with Pen and Pencil* (London: The Religious Tract Society, 1893), 28.
58. Kelley Wilder, *Photography and Science* (Chicago: University of Chicago, 2009), 10-11.
59. "Lecture on Iceland," *Birmingham Daily Post*, March 8, 1892.
60. Wilder, *Photography and Science*, 124.
61. Guðmundsson, Hannesdóttir, and Björnsson, "Post-Little Ice Age Volume Loss of Kotárjökull Glacier," 108.
62. Frank Ponzi, *Islands Howells/Howell's Iceland: 1890-1901* (Mosfellsbær, Iceland: Brennholt, 2004), 177.
63. Mark Klett, Byron Wolfe, and Rebecca Solnit, *Yosemite in Time: Ice Ages, Tree Clocks, Ghost Rivers* (San Antonio: Trinity University Press, 2008), x.

64. Jordan Bear and Kate Palmer Albers, *Before-and-After Photography: Histories and Contexts* (New York, NY: Bloomsbury, 2017), 2.
65. Kimmerer, "Weaving Traditional and Ecological Knowledge," 433.
66. Shepherd, *The Living Mountain,* 107–8.
67. M Jackson, *The Secret Lives of Glaciers* (Brattleboro, VT: Green Writers Press, 2019), chapter 7.
68. Lauren E. Cagle and Denise Tillery, "Climate Change Research Across Disciplines: The Value and Uses of Multidisciplinary Research Reviews for Technical Communication," *Technical Communication Quarterly* 25, no. 2 (2015): 160, https://doi.org/10.1080/10572252.2015.1001296.
69. Gregory N. Bratman et al., "Nature Experience Reduces Rumination and Subgenual Prefrontal Cortex Activation," *Proceedings of the National Academy of the Sciences of the United States of America* 112, no. 28 (2015), https://doi.org/10.1073/pnas.1510459112; Eugenia C. South et al., "Effect of Greening Vacant Land on Mental Health of Community-Dwelling Adults: A Cluster Randomized Trial," *Journal of the American Medical Association* 1, no. 3 (2018), doi:10.1001/jamanetworkopen.2018.0298; Eugenia C. South et al., "Neighborhood Blight, Stress, and Health: A Walking Trial of Urban Greening and Ambulatory Heart Rate," *American Journal of Public Heath* 105, no. 5 (2015), doi:10.2105/AJPH.2014.302526.
70. Natasha Seegert, "Play of Sniffication: Coyotes Sing in the Margins," *Philosophy & Rhetoric* 47, no. 2 (2014): 164. https://doi.org/10.5325/philrhet.47.2.0158.
71. Joshua Trey Barnett, "Rhetoric for Earthly Coexistence: Imagining an Ecocentric Rhetoric," *Rhetoric & Public Affairs* 24, no. 1–2 (2021): 370.
72. Shepherd, *The Living Mountain,* 107–8.

Bibliography

Abdullahi, Ali Arazeem. "Trends and Challenges of Traditional Medicine in Africa." *African Journal of Traditional, Complementary, and Alternative Medicines* 8, no. 5 (2011): 115–23. https://doi:10.4314/ajtcam.v8i5S.5

Adams, Heather B. "Decolonizing Projects: Creating Pluriversal Possibilities/ This Fractured Pedagogy: Listening and Learning from My Best Teacher." *Rhetoric Review* 38, no. 1 (January 2019): 18–22. https://doi.org/10.1080/07350198.2019.1549402.

Adefolaju, Toyin. "The Dynamics and Changing Structure of Traditional Healing System in Nigeria." *International Journal of Health Research* 4, no. 2 (2011).

Adeoye, C. L. *Asa Ati Ise Yoruba*. Ibadan: Oxford, 1979.

Aderibigbe, I. S. "The Traditional Healing System among the Yoruba." In *Traditional and Modern Health Systems in Nigeria AWP Trenton and Asmara*, edited by Toylin Falola and Matthew H. Heaton, 365–80. Trenton, NJ: Africa World Press, 2006.

Afisi, Oseni Taiwo. "Is African Science True Science? Reflections on the Methods of African Science." *Filosofia Theoretica* 5, no. 1 (2016). https://doi.org/10.4314/ft.v5i1.5.

Agboka, Godwin Y. "Decolonial Methodologies: Social Justice Perspectives in Intercultural Technical Communication Research." *Journal of Technical Writing and Communication* 44, no. 3 (2014): 297–327.

Aimers, James J., and Prudence M. Rice. "Astronomy, Ritual, and the Interpretation of Maya 'E-Group' Architectural Assemblages." *Ancient Mesoamerica* 17, no. 1 (2006): 79–96.

Al-Khalili, Jim. *The House of Wisdom: How Arabic Science Saved Ancient Knowledge and Gave Us the Renaissance*. London: Penguin, 2011.

Allen, Chadwick. *Blood Narrative Indigenous Identity in American Indian and Maori Literary and Activist Texts*. Durham, NC: Duke University Press, 2002.

Allen, J. Romilly. *Celtic Art in Pagan and Christian Times*. London: Methuen & Co., 1904.

Altman, Jon Charles, and Peter Whitehead. *Caring for Country and Sustainable Indigenous Development: Opportunities, Constraints and Innovation.* Canberra: Australian National University Press, 2003.

Anderson, Kay, and Colin Perrin. "'Removed from Nature' The Modern Idea of Human Exceptionality." *Environmental Humanities* 10, no. 2 (2018): 447–72.

Anti-Defamation League. "Celtic Cross." Accessed December 12, 2022. https://www.adl.org/education/references/hate-symbols/celtic-cross.

Appadurai, Arjun. "Grassroots Globalization and the Research Imagination." *Public culture* 12, no. 1 (2000): 1–19.

Apuzzo, Matt, and Selam Gebrekidan. "For Covid-19 Vaccines, Some Are Too Rich—And Too Poor." *New York Times*, December 28, 2020.

"AQI Levels and Meanings Released by the U.S. Embassy in Beijing." *United States Embassy to China*, 2011. Accessed November 21, 2020. https://china.usembassy-china.org.cn/embassy-consulates/beijing/air-quality-monitor/?_ga=2.149520370.559911029.1579965848-1378292718.1579965848.

Arola, Kristin L. "Indigenous Interfaces." In *Social Writing/Social Media: Publics, Presentations, and Pedagogies*, edited by Kristin L. Arola, 211–26. Fort Collins, CO: WAC, 2017. https://wac.colostate.edu/docs/books/social/chapter11.pdf.

Ascher, Marcia, and Robert Ascher. *Code of the Quipu: A Study in Media, Mathematics, and Culture.* Ann Arbor: University of Michigan Press, 1981.

Ásmundarson, Valdimar. *Egils saga Skallagrímssonar.* Reykjavík: Kostnaðarmaður, Sigurður Kristjánsson, 1892.

Attenbrow, Val. *Sydney's Aboriginal Past: Investigating the Archaeological and Historical Records.* Sydney: University of New South Wales Press, 2010.

Baake, Ken. *Metaphor and Knowledge: The Challenges of Writing Science.* Albany, NY: SUNY Press, 2003.

Babalola, Abidemi Babatunde, Laure Dussubieux, Susan Keech McIntosh, and Thilo Rehren. "Chemical Analysis of Glass Beads from Igbo Olokun, Ile-Ife (SW Nigeria): New Light on Raw Materials, Production, and Interregional Interactions." *Journal of Archaeological Science* 90 (2018): 92–105.

Baca, Damián and Victor Villanueva, eds. *Rhetorics of the Americas: 3114 BCE to 2012 CE.* New York: Palgrave Macmillan, 2010.

Bahn, Paul, and John Flenley. *Easter Island, Earth Island.* 4th ed. Lanham, ML: Rowman & Littlefield, 2017 (1992).

Bakar, Osman. *Tawhid and Science: Essays on the History and Philosophy of Islamic Science.* Penang, Malaysia: Secretariat for Islamic Philosophy and Science, 1991.

Bala, Arun, and George Gheverghese Joseph. "Indigenous Knowledge and Western Science: The Possibility of Dialogue." *Race & Class* 49, no. 1 (2007): 39–61.

Banner, Stuart. "Why Terra Nullius-Anthropology and Property Law in Early Australia." *Law & History Review* 23 (2005): 95–131.

Barbour, Wayne, and Christine Schlesinger. "Who's the Boss? Post-Colonialism, Ecological Research and Conservation Management on Australian Indige-

nous Lands." *Ecological Management & Restoration* 13, no. 1 (2012): 36–41. https://doi.org/10.1111/j.1442-8903.2011.00632.x.
Barnett, Joshua Trey. "Rhetoric for Earthly Coexistence: Imagining an Ecocentric Rhetoric." *Rhetoric & Public Affairs* 24, no. 1–2 (2021): 365–78.
Barthel, Thomas S. *The Eighth Land: The Polynesian Settlement of Easter Island*. Honolulu: University of Hawai'i Press, 1978.
Bateson, Gregory. *Steps to an Ecology of Mind*. New York: Ballantine Books, 1972.
Bear, Jordan, and Kate Palmer Albers, *Before-and-After Photography: Histories and Contexts*. New York: Bloomsbury, 2017.
Beck, Ulrich. *World Risk Society*. Malden, MA: Polity Press, 1999.
Beck, Ulrich. *Risk Society: Towards a New Modernity*. London: Sage Publications, 1992.
Beck, Ulrich, Scott Lash, and Brian Wynne. *Risk Society: Towards a New Modernity*. London: Sage, 1992.
Bellwood, Peter S. *The Polynesians: Prehistory of an Island People*. London: Thames and Hudson, 1978.
Berkes, Fikret. *Sacred Ecology*. New York: Routledge, 2018.
Berndt, Ronald Murray, and Catherine Helen Berndt. *The World of the First Australians*. Sydney: Lansdowne, 1981.
Berry, Wendell. "Life is a Miracle." *Communio: International Catholic Review* 27 (2000): 83–97.
Birch, Kean and David Tyfield. "Theorizing the Bioeconomy: Biovalue, Biocapital, Bioeconomics or . . . What?" *Science, Technology, and Human Values* 38, no. 3 (2012): 299–327. http://doi.org/10.1177/0162243912442398.
Björnsson, Helgi, and Finnur Pálsson. "Icelandic Glaciers." *Jökull* no. 58 (2008): 265–86.
Björnsson, Helgi. *The Glaciers of Iceland: A Historical, Cultural, and Scientific Overview*. Translated by Julian Meldon D'Arcy. Durham, NC: Atlantic Press, 2017.
Black, Max. *Models and Metaphors*. Ithaca, NY: Cornell University Press, 2019.
Boer, Richard Constant, ed. *Grettis saga Asmundarsonar*. Magdeburg, Germany: Niemeyer, 1900.
Bohensky, Erin L., James R. A. Butler, and Jocelyn Davies. "Integrating Indigenous Ecological Knowledge and Science in Natural Resource Management: Perspectives from Australia." *Ecology and Society* 18, no. 3 (2013). https://www.jstor.org/stable/26269334.
Bonta, Mark, et al., "Intentional Fire-Spreading by 'Firehawk' Raptors in Northern Australia." *Journal of Ethnobiology* 37, no. 4 (2017): 700–18.
Borokini, Temitope I., and Ibrahim O. Lawal. "Traditional Medicine Practices among the Yoruba People of Nigeria : A Historical Perspective." *Journal of Medicinal Plants Studies* 2, no. 6 (2014).
Bowman, D. M. J. S. "The Impact of Aboriginal landscape Burning on the Australian Biota." *New Phytologist* 140, no. 3 (1998): 385–410. https://doi.org/10.1111/j.1469-8137.1998.00289.x.

Bratman, Gregory N., et al. "Nature Experience Reduces Rumination and Subgenual Prefrontal Cortex Activation." *Proceedings of the National Academy of the Sciences of the United States of America* 112, no. 28 (2015). https://doi.org/10.1073/pnas.1510459112.

Brown, Theodore L. *Making Truth: Metaphor in Science*. Chicago: University of Illinois Press, 2003.

Buranyi, Stephen. "Big Pharma is Fooling Us." *New York Times*, December 17, 2020.

Burke, Kenneth. *A Rhetoric of Motives*. New York: Prentice Hall, 1950.

Burke, Kenneth. *Language As Symbolic Action: Essays on Life, Literature, and Method*. Berkeley: University of California Press, 1966.

Cagle, Lauren E., and Denise Tillery. "Climate Change Research Across Disciplines: The Value and Uses of Multidisciplinary Research Reviews for Technical Communication." *Technical Communication Quarterly* 25, no. 2 (2015): 147–63. https://doi.org/10.1080/10572252.2015.1001296.

Calder, Muffy, Claire Craig, Dave Culley, Richard de Cani, Christl A. Donnelly, Rowan Douglas, Bruce Edmonds, et al. "Computational Modelling for Decision-Making: Where, Why, What, Who and How." *Royal Society Open Science* 5, no. 6 (2018): 172096. https://doi.org/10.1098/rsos.172096.

Callon, Michel, Pierre Lascoumes, and Yannick Barthe. *Acting in an Uncertain World: An Essay on Technical Democracy (inside Technology)*. Cambridge, MA: MIT Press, 2009.

Cane, Scott. *Pila Nguru: The Spinifex People*. Sydney: Fremantle Press, 2002.

Carpentier, L., T. Prazuck, F. Vincent-Ballereau, L. T. Ouedraogo, and C. Lafaix. "Choice of Traditional or Modern Treatment in West Burkina Faso." *World Health Forum* 16, no. 2 (1995): 198–202.

"Celtic Art in Britain, Ireland." *Encyclopedia of Irish and Celtic Art*. Accessed December 12, 2022. http://www.visual-arts-cork.com/cultural-history-of-ireland/celtic-art-in-britain-ireland.htm.

"Celtic Interlace Designs." *Encyclopedia of Irish and Celtic Art*. Accessed December 12, 2022. http://www.visual-arts-cork.com/cultural-history-of-ireland/celtic-interlace-designs.htm.

Charnley, Susan, A. Paige Fischer, and Eric T. Jones, "Integrating Traditional and Local Ecological Knowledge into Forest Biodiversity Conservation in the Pacific Northwest." *Forest Ecology and Management*, 246 (2007): 14–28. https://doi.org/10.1016/j.foreco.2007.03.047.

Clahchischiligi, Sunnie R. "Navajo Elders: Alone, without Food, in Despair." *Searchlight NM*, August 6, 2020. https://searchlightnm.org/navajo-elders-alone-without-food-in-despair/.

Clarkson, Chris, et al. "Human Occupation of Northern Australia by 65,000 Years Ago." *Nature* 547, no. 7663 (2017): 306–10. https://doi.org/10.1038/nature22968.

Clary-Lemon, Jennifer. *Planting the Anthropocene: Rhetorics of Natureculture*. Louisville, CO: Utah State University, 2019.

Cobos, Casie, Gabriela Raquel Ríos, Donnie Johnson Sackey, Jennifer Sano-Franchini, and Angela M. Haas. "Interfacing Cultural Rhetorics: A History and a Call." *Rhetoric Review* 37, no. 2 (2018): 139–54. https://doi.org/10.10 80/07350198.2018.1424470.
Colley, Sarah, and Val Attenbrow. "Does Technology Make a Difference? Aboriginal and Colonial Fishing in Port Jackson, New South Wales." *Archaeology in Oceania* 47, no. 2 (2012): 69–77.
Collis, John. "The Sheffield Origins of Celtic Art." In *Rethinking Celtic Art*, edited by Duncan Garrow, Chris Gosden, and J.D. Hill, 21–27. Oxford and Philadelphia: Oxbow Books, 2008.
Condit, Celeste M. "How the Public Understands Genetics: Non-Deterministic and Non-Discriminatory Interpretations of the 'Blueprint' Metaphor." *Public Understanding of Science* 8, no. 3 (1999): 169–80. https://doi.org/10.1088/ 0963-6625/8/3/302. https://doi.org/10.1088/0963-6625/8/3/302.
Conger, Krista. "Researchers Awarded $31 Million for Clinical Trials to Treat Stroke, Heart Failure, Brain Cancer." Stanford Medicine News Center, *Stanford Medicine*. September 9, 2021. https://med.stanford.edu/news/all-news/2021/09/grants-for-stem-cell-clinical-trials.html.
Connor, Jessica, and Nick Ward. "Celtic Knot Theory." Undergraduate Student Project, University of Edinburgh, 2012.
Cook, Gary D, Sue Jackson, and Richard J Williams. "A Revolution in Northern Australian Fire Management: Recognition Of Indigenous Knowledge, Practice And Management." In *Flammable Australia*, edited by Ross A Bradstock, A Malcolm Gill, and Richard J. Williams, 293–305. Collingwood: CSIRO Publishing, 2012.
Cook, Harold. "Sciences and Economies in the Scientific Revolution: Concepts, Materials, and Commensurable Fragments." *OSIRIS* 33 (2018): 25–44. http:// doi.org/10.1086/699171.
Coombs, Herbert Cole, Barrie Graham Dexter, and Lester Richard Hiatt. "The Outstation Movement in Aboriginal Australia." *Australian Institute of Aboriginal Studies Newsletter* 14 (1980): 16–23.
Corbett, Edward P. J. and Robert J. Connors. *Style and Statement*. New York: Oxford University Press, 1999.
Cox, Paul Alan, and Sandra Anne Banack, eds. *Islands, Plants and Polynesians: An Introduction to Polynesian Ethnobotany*. Portland, OR: Dioscorides Press, 1991.
Cristino, Claudio, Patricia Vargas, and Roberto Izaurieta. *Atlas arqueólogica de Isla de Pascua*. Santiago: Editorial Universitaria, 1981.
Cromwell, Peter R. "Celtic Knotwork: Mathematical Art." *The Mathematical Intelligencer* 15, no. 1 (1993): 36–47.
Cronon, William. "The Trouble with Wilderness, or Getting Back to the Wrong Nature." *Environmental History* 1, no. 1 (1996): 24. http://www.jstor.org/stable/ 3985059.

Crowe, Andrew. *Pathway of the Birds: The Voyaging Achievements of Māori and Their Polynesian Ancestors*. Honolulu: University of Hawai'i Press, 2018.
Cultural Rhetorics Theory Lab. "Our Story Begins Here: Constellating Cultural-Rhetorics." *Enculturation* 21, no. 1 (October 2015). https://enculturation.net/our-story-begins-here.
Cunsolo, Ashlee, and Neville R. Ellis. "Ecological Grief as a Mental Health Response to Climate Change-Related Loss." *Climate Change* 8, no. 4 (2018): 275–81.
Cushman, Ellen. "Translingual and Decolonial Approaches to Meaning Making." *College English* 78, no. 3 (2016): 234–42.
Cutler, David, and Francesca Dominici. "A Breath of Bad Air: Cost of the Trump Environmental Agenda May Lead to 80 000 Extra Deaths Per Decade." *Jama* 319, no. 22 (2018): 2261–62.
D'Ambrosio, Ubiratan. *Mathematics across Cultures: The History of Non-Western Mathematics*. Vol. 2. Berlin: Springer Science & Business Media, 2001.
Dangor, Suleman. "Islamization of Disciplines: Towards an Indigenous Educational System." *Educational Philosophy and Theory* 37, no. 4 (2005): 519–31. https://doi.org/10.1111/j.1469-5812.2005.00138.x.
Davis, Lisa A., and Sara Bishop. "The Celtic Knot Project: A Holistic Nursing Intervention Teaching Strategy." *International Journal for Human Caring* 7, no. 3 (October 2003): 8–12. https://doi.org/10.20467/1091-5710.7.3.9.
Dawson, G., B. V. Lightman, M. Elshakry, and S. Sivasundaram. *Victorian Science and Literature: Science, Race and Imperialism*. Pickering & Chatto, 2012. https://books.google.com/books?id=wpl1mwEACAAJ.
De Burgh, Hugo, and Rong Zeng. "Environment Correspondents in China in Their Own Words: Their Perceptions of Their Role and the Possible Consequences of Their Journalism." *Journalism* 13, no. 8 (2012): 1004–23.
De Pryck, Kari, and François Gemenne. "The Denier-in-Chief: Climate Change, Science and the Election of Donald J. Trump." *Law and Critique* 28, no. 2 (2017): 119–26.
Delhon, Claire, and Catherine Orliac. "The Vanished Palm Trees of Easter Island: New Radiocarbon and Phytolith Data." In *VII International Conference on Easter Island and the Pacific Aug 2007*. Visby: Gotland University Press, 2010.
Deloria, Vine. *We Talk, You Listen*. Lincoln: University of Nebraska Press, 1970.
Denetclaw, Pauly. "Young Signmakers: Whatever You Choose to Call It, Keep It Home!" *The Navajo Times*, April 9, 2020. https://navajotimes.com/coronavirus-updates/young-signmakers-whatever-you-choose-to-call-it-keep-it-home/.
Deng, Xiaobei, Fang Zhang, Wei Rui, Fang Long, Lijuan Wang, Zhaohan Feng, Deliang Chen, and Wenjun Ding. "Pm2. 5-Induced Oxidative Stress Triggers Autophagy in Human Lung Epithelial A549 Cells." *Toxicology in Vitro* 27, no. 6 (2013): 1762–70.
Diamond, Jared. *Collapse: How Societies Choose to Fail or Succeed*. Revised Edition. New York: Penguin, 2011 (2005).

Ding, Huiling and Zhang, Jingwen. "Imagining Health Risks: Fear, Fate, Death, and Family in Chinese and American Online Discussion Forums About Hiv/Aids." In *Imagining China: Rhetorics of Nationalism in an Age of Globalization*, edited by Stephen John Hartnett, Lisa B. Keränen, and Donovan Conley, 235–70. Lansing: Michigan State University Press, 2017.

Ding, Huiling. *Rhetoric of a Global Epidemic: Intercultural and Intracultural Professional Communication About Sars*. PhD diss., Purdue University, 2007.

Ding, Huiling. *Rhetoric of a Global Epidemic: Transcultural Communication About Sars*. Carbondale: Southern Illinois University Press, 2014.

Ding, Huiling. "Rhetorics of Alternative Media in an Emerging Epidemic: Sars, Censorship, and Extra-Institutional Risk Communication." *Technical Communication Quarterly* 18, no. 4 (2009): 327–50.

Ding, Huiling "Transcultural Risk Communication and Viral Discourses: Grassroots Movements to Manage Global Risks of H1n1 Flu Pandemic." *Technical Communication Quarterly* 22, no. 2 (2013): 126–49.

Doran, B. R. "Mathematical Sophistication of the Insular Celts—Spirals, Symmetries, and Knots as a Window onto Their World View." *Proceedings of the Harvard Celtic Colloquium* 15 (1995): 258–89.

Dransfield, John, John Flenley, Sarah M. King, D. D. Harkness and Sergio Rapu. "A Recently Extinct Palm from Easter Island." *Nature* 312 (1984): 750–752.

Druschke, Caroline Gottschalk, and Kristen C. Hychka. "Manager Perspectives on Communication and Public Engagement in Ecological Restoration Project Success." *Ecology and Society* 20, no. 1 (2015): Article 58.

Du, Shaozhong. "We Are Weak in Communicating Air Quality." Weibo, November 1, 2011. Accessed November 21, 2020. https://www.weibo.com/dushaozhong?is_hot=1.

Du, Shaozhong. "On the Differences of Air Quality Ratings." Weibo, October 31, 2011. Accessed November 21, 2020. https://www.weibo.com/dushaozhong?is_hot=1.

Dunaway, Finis. *Seeing Green: The Use and Abuse of American Environmental Images*. Chicago: University of Chicago Press, 2015.

Eades, Jeremy Seymour. *The Yoruba Today*. Cambridge: Cambridge UP, 1980.

Edbauer, Jenny. "Unframing Models of Public Distribution: From Rhetorical Situation to Rhetorical Ecologies." *Rhetoric Society Quarterly* 35, no. 4 (2005): 5–24.

Edwards, Alexandra, Barthélémy d'Ans and Edmundo Edwards. "Consolidation of the Rapanui Astronomy Concept Inventory and Re-appraisal of Applied Astronomic Observation at Papa Ui Hetu'u Rapa Nui." *Mediterranean Archaeology and Archaeometry* 18, no. 4 (2018): 139–47.

Edwards, Edmundo, and Juan Antonio Belmonte. "Megalithic Astronomy of Easter Island: A Reassessment." *Journal of the History of Astronomy* 35 (2004): 421–433.

Edwards, Edmundo, and Alexandra Edwards. *When the Universe Was an Island: Exploring the Cultural and Spiritual Cosmos of Ancient Rapa Nui.* Easter Island: Hangaroa Press, 2013.

Edwards, Kevin J., Ian Thomas Lawson, Egill Erlendsson, and Andrew Dugmore. "Landscapes of Contrast in Viking Age Iceland and the Faroe Islands." *Landscapes* 6, no. 2 (2005): 63–81. http://doi.org/10.1179/lan.2005.6.2.63.

Ehrlich, Paul. *The Population Bomb.* New York: Ballantine Books, 1968.

El-Bushra, El-Sayed, and M. M. Muhammadain. "Perspectives on the Contribution of Arabs and Muslims to Geography." *GeoJournal* 26, no. 2 (1992): 157–66.

Elder, Bruce. *Blood on the Wattle: Massacres and Maltreatment of Aboriginal Australians since 1788.* Sydney: New Holland Press, 2003.

Elshakry, Marwa S. "Knowledge in Motion: The Cultural Politics of Modern Science Translations in Arabic." *Isis* 99, no. 4 (2008): 701–30.

Emeagwali, G., and G. J. S. Dei. *African Indigenous Knowledge and the Disciplines.* Rotterdam: SensePublishers, 2014.

Endres, Danielle. "Engaging the Nexus of Environmental Rhetoric and Indigenous Rights." *Spectra* 55, no. 1 (March 2019): 24–25

Englert, Sebastian. *La Tierra de Hotu Matu'a: historia y etnología de la Isla de Pascua.* 9th Edition. Santiago: Editorial Universitaria, 2004 (1948).

Ens, Emilie J., et al. "Indigenous Biocultural Knowledge in Ecosystem Science and Management: Review and Insight from Australia." *Biological Conservation* 181 (2015): 133–49. https://doi.org/10.1016/j.biocon.2014.11.008.

Enos, Richard Leo. "The Archaeology of Women in Rhetoric: Rhetorical Sequencing as a Research Method for Historical Scholarship." *Rhetoric Society Quarterly* 32, no. 1 (January 1, 2002): 65–79. https://doi.org/10.1080/02773940209391221.

Ens, Emilie, Mitchell L. Scott, Yugul Mangi Rangers, Craig Moritz, and Rebecca Pirzl. "Putting Indigenous Conservation Policy into Practice Delivers Biodiversity and Cultural Benefits." *Biodiversity and Conservation* 25, no. 14 (2016): 2889–906.

Esteban, César. "Some Notes on Orientations of Prehistoric Stone Monuments in Western Polynesia and Micronesia." *Archæoastronomy: The Journal of Astronomy in Culture.* XVII (2002–2003): 31–47.

Etzkowitz, Henry, and Loet Leydesdorff. "The Dynamics of Innovation: From National Systems and 'Mode 2' to a Triple Helix of University–Industry–Government Relations." *Research Policy* 29, no. 2 (2000): 109–23.

Fache, Elodie. "Caring for Country, a Form of Bureaucratic Participation. Conservation, Development, and Neoliberalism in Indigenous Australia." *Anthropological Forum* 24, no. 3 (2014): 267–86.

Farley, Julia, and Fraser Hunter, eds. *Celts: Art and Identity.* London: British Museum Press: National Museums, Scotland, 2015.

Faruqi, Yasmeen Mahnaz. "Contributions of Islamic Scholars to the Scientific Enterprise." *International Education Journal* 7, no. 4 (2006): 391–99.

Fedorenko, Irina, and Yixian Sun. "Microblogging-Based Civic Participation on Environment in China: A Case Study of the Pm 2.5 Campaign." *VOLUNTAS: International Journal of Voluntary and Nonprofit Organizations* 27, no. 5 (2016): 2077–105.

Feng, Jie, and Lu Zongshu, "I Gauge the Air Quality for My Motherland," *Southern Weekly*, October 28, 2011. http://www.infzm.com/content/64281.

Finney, Ben. "Voyaging Canoes and the Settlement of Polynesia." *Science* 196 (1977): 1277–85.

Finney, Ben. "The Impact of Late Holocene Climate Change on Polynesia." *Rapa Nui Journal* 8, no. 1 (1994): 13–14.

Finney, Ben. "Tracking Polynesian Seafarers." *Science* 317, no. 5846 (28 Sept. 2007): 1873–74.

Fischer, Steven Roger. "Rapanui's Tu'u ko Iho versus Mangareva's 'Atu Motua: Evidence for Multiple Reanalysis and Replacement in Rapanui Settlement Traditions, Easter Island." *The Journal of Pacific History* 29, no. 1 (1994): 3–18.

Fisher, G., and B. Mellor. "On the Topology of Celtic Knot Designs." *Proceedings of the Bridges Conference: Mathematical Connections in Art, Music, and Science*, 2004. http://www.mi.sanu.ac.rs/vismath/fisher/index.html.

Flenley, John R. "The Palaeoecology of Rapa Nui, and its Ecological Disaster." *Easter Island Studies: Contributions to the History of Rapanui in Memory of William T. Mulloy*, edited by S. R. Fischer, 27–45. Oxford: Oxford Books, 1993.

Flenley, John R., and Paul Bahn. *The Enigmas of Easter Island*. New York: Oxford University Press, 2002.

Flenley, John R., and Paul Bahn. "Conflicting Views of Easter Island." *Rapa Nui Journal* 21, no. 1 (2007): 11–13.

Flores, Lisa. "Advancing a Decolonial Rhetoric." *Advances in the History of Rhetoric* 21, no. 3 (2018): 320–22.

Francis, Mark. "Social Darwinism and the Construction of Institutionalised Racism in Australia." *Journal of Australian Studies* 20, no. 50–51 (1996): 90–105.

Frankenfeld, Philip J. "Technological Citizenship: A Normative Framework for Risk Studies." *Science, Technology, & Human Values* 17, no. 4 (1992): 462–65.

Fraser, Nancy. "Expropriation and Exploitation in Racialized Capitalism: A Reply to Michael Dawson." *Critical Historical Studies* 3, no. 1(Spring 2016): 163–78.

Gadgil, M., F. Berkes, and C. Folke. "Indigenous Knowledge for Biodiversity Conservation." *Ambio* 22 (1993): 151–56.

Gammage, B. *The Biggest Estate on Earth: How Aborigines Made Australia*. Melbourne: Allen and Unwin, 2011.

García, Romeo, and Damián Baca. "Rhetorics Elsewhere and Otherwise: Contested Modernities, Decolonial Visions." Champaign, IL: National Council of Teachers of English, 2019.

Genz, Joseph H. "Resolving Ambivalence in Marshallese Navigation: Relearning, Reinterpreting, and Reviving the 'Stick Chart' Wave Models." *Structure and Dynamics* 9, no. 1 (2016): 8–40.

Gerritson, R. *Australia and the Origins of Agriculture*. Oxford: Archaeopress, 2008.

Ghosh, Arunabh. "Lies, Damned Lies, and (Bourgeois) Statistics: Ascertaining Social Fact in Midcentury China and the Soviet Union." *OSIRIS* 33 (2018): 149–68. http://doi.org/10.1086/699237.

Giles, Melanie. "Seeing Red: The Aesthetics of Martial Objects in the British and Irish Iron Age." In *Rethinking Celtic Art*, edited by Duncan Garrow, Chris Gosden, and J. D. Hill, 59–77. Oxford and Philadelphia: Oxbow Books, 2008.

Giridharadas. Anand. "Does the Climate Movement Need a Makeover?" *The Ink*, August 3, 2021. https://the.ink/p/does-the-climate-movement-need-a.

Good, Charles M., John M. Hunter, Selig H. Katz, and Sydney S. Katz. "The Interface of Dual Systems of Health Care in the Developing World: Toward Health Policy Initiatives in Africa." *Social Science & Medicine. Part D: Medical Geography* 13, no. 3 (1979): 141–54.

Goodnight, G. Thomas. "The Personal, Technical, and Public Spheres of Argument: A Speculative Inquiry into the Art of Public Deliberation." *The Journal of the American Forensic Association* 18, no. 4 (1982): 214–27.

Goodnight, G. Thomas. "The Personal, Technical, and Public Spheres of Argument: A Speculative Inquiry into the Art of Public Deliberation." *Argumentation and Advocacy* 48, no. 4 (2012): 198–210.

Goodnight, G. Thomas. "The Personal, Technical, and Public Spheres: A Note on 21St Century Critical Communication Inquiry," *Argumentation and Advocacy: Special Issue: Spheres of Argument: 30 Years of Goodnight's Influence* 48, no. 4 (2012): 257–67.

González-Ferrán, Óscar, et al. *Geología del complejo volcánico Isla de Pascua Rapa Nui*. Santiago: Centro de Estudios Volcanológicos, 2004.

Gosden, Chris, and J. D. Hill. "Introduction: Re-Integrating 'Celtic' Art." In *Rethinking Celtic Art*, edited by Duncan Garrow, Chris Gosden, and J. D. Hill, 1–14. Oxford and Philadelphia: Oxbow Books, 2008.

Goss, Christopher H., Stacey A. Newsom, Jonathan S. Schildcrout, Lianne Sheppard, and Joel D. Kaufman. "Effect of Ambient Air Pollution on Pulmonary Exacerbations and Lung Function in Cystic Fibrosis." *American Journal of Respiratory and Critical Care Medicine* 169, no. 7 (2004): 816–21.

Graham, S. Scott. *The Politics of Pain Medicine: A Rhetorical-Ontological Inquiry*. Chicago: University of Chicago Press, 2015.

Grebmeier, Jacqueline M., James E. Overland, Sue. E. Moore, Ed V. Farley, Eddy C. Carmack, Lee W. Cooper, Karen E. Frey, John H. Helle, Fiona A. McLaughlin, S. Lyn McNutt. "A Major Ecosystem Shift in the Northern Bering Sea." *Science* 311 (2006): 1461–64.

Green Beagle. "No Car Day," Weibo, September 22, 2011. https://www.weibo.com/greenbeagle/profile?s=6cm7D0.
Green Beagle. "Blue Sky Created by Standards." Weibo, October 7, 2011. https://www.weibo.com/greenbeagle/profile?s=6cm7D0.
"Green Water and Blue Sky: Overview of the Ecological Civilization Efforts after the 18th People's Congress." November 11, 2013. http://www.gov.cn/jrzg/2013-11/11/content_2525087.htm.
Gries, Laurie E. "Iconographic Tracking: A Digital Research Method for Visual Rhetoric and Circulation Studies." *Computers and Composition* 30, no. 4 (December 1, 2013): 332–48. https://doi.org/10.1016/j.compcom.2013.10.006.
Gross, Alan G. "Toward a Theory of Verbal-Visual Interaction: The Example of Lavoisier." *Rhetoric Society Quarterly* 39, no. 2 (2009): 147–69.
Gross, Alan G., and William M Keith. *Rhetorical Hermeneutics: Invention and Interpretation in the Age of Science*. Albany, NY: SUNY Press, 1997.
Guðmundsson, Snævarr, Hrafnhildur Hannesdóttir, and Helgi Björnsson. "Post-Little Ice Age Volume Loss of Kotárjökull Glacier, SE-Iceland, Derived from Historical Photography." *Jökull* 62 (2012): 97–110.
Guerra, Francisco. "Aztec Science and Technology." *History of Science* 8, no. 1 (1969): 32–52.
Guo, Baogang. "Political Legitimacy and China's Transition." *Journal of Chinese Political Science* 8, no. 1–2 (2003): 1–25.
Haas, Angela M. "Race, Rhetoric, and Technology: A Case Study of Decolonial Technical Communication Theory, Methodology, and Pedagogy." *Journal of Business and Technical Communication* 26, no. 3 (2012): 277-310.
Haas, Angela M., "Wampum as Hypertext: An American Indian Intellectual Tradition of Multimedia Theory and Practice." *Studies in American Indian Literatures* 19, no. 4 (December 2007): 77–100. https://www.jstor.org/stable/20737390.
Hagelberg, Erika. "Genetic Affinities of the Rapanui." In *Skeletal Biology of the Ancient Rapa Nui*, edited by G. W. Gill and V. Stefan, 182–201. Cambridge: Cambridge University Press, 2016.
Harding, Sandra, ed. *The "Racial" Economy of Science: Toward a Democratic Future*. Bloomington: Indiana University Press, 1993.
Harris, Randy Allen. *Landmark Essays on Rhetoric of Science Case Studies*. Mahwah, NJ: Lawrence Erlbaum Associates, 1997.
Harris, Randy Allen. *Rhetoric and Incommensurability*. Chicago: Parlor Press, 2005.
Hastrup, Kirsten. "Icelandic Topography and The Sense of Identity." In *Nordic Landscapes*, edited by Kenneth Olwig and Michael Jones, 53–76. Minneapolis: University of Minnesota Press, 2008.
Helmreich, Stefan. "Blue-green Capital, Biotechnological Circulation and an Oceanic Imaginary: A Critique of Biopolitical Economy." *BioSocieties*. 2 (2007): 287–302. http://doi.org/10.1017/S1745855207005753.

Harte, Jeff. "Ancient Celtic Knots Inspire Scientific Breakthrough." *The Irish Times*, May 21, 2013. https://www.irishtimes.com/news/science/ancient-celtic-knots-inspire-scientific-breakthrough-1.1401644.

Hather, Jon G. "The Archaeobotany of Subsistence in the Pacific," *World Archaeology* 24, no. 1 (June 1992): 70–81.

Hawk, Byron. *A Counter-History of Composition: Toward Methodologies of Complexity*. 1st ed. Pittsburgh, PA: University of Pittsburgh Press, 2007.

Henderson, Ebenezer. *Iceland; Or, the Journal of a Residence in That Island, During the Years 1814 and 1815*. Edinburgh, Scotland: Oliphant, Waugh, and Innes.

Heyerdahl, Thor. *The Kon-Tiki Expedition*. London: Allen & Unwin, 1950.

Heyerdahl, Thor. *American Indians in the Pacific*. London: Allen & Unwin, 1952.

Heyerdahl, Thor. *Aku-Aku: The Secret of Easter Island*. London: Allen & Unwin, 1958.

Heyerdahl, Thor, and Edwin Ferdon, eds. *Archaeology of Easter Island*. Chicago: Rand McNally, 1961.

Hill, Rosemary, Adelaide Baird, and David Buchanan. "Aborigines and Fire in the Wet Tropics of Queensland, Australia: Ecosystem Management Across Cultures." *Society & Natural Resources* 12, no. 3 (1999): 205–23.

Hill, Rosemary, et al. *Yalanji-Warranga Kaban: Yalanji People of the Rainforest Fire Management Book*. Queensland, Australia: Little Ramsay Press, 2004.

"Historical Archaeology in Africa: Representation, Social Memory, and Oral Traditions." *Choice Reviews Online* 44, no. 12 (2007). https://doi.org/10.5860/choice.44-6922.

Ho, Peter, and Richard Edmonds. *China's Embedded Activism: Opportunities and Constraints of a Social Movement*. London: Routledge, 2007.

Holbig, Heike, and Bruce Gilley. "Reclaiming Legitimacy in China." *Politics & Policy* 38, no. 3 (2010): 395–422.

Holton, Graham E. L. "Heyerdahl's Kon Tiki Theory and the Denial of the Indigenous Past." *Anthropological Forum* 14, no. 2 (2004), 163–181.

Horrebow, Niels. *Tilforladelige Efterretninger om Island*. Copenhagen, 1752.

Horswell, Michael J. *Decolonizing the Sodomite: Queer Tropes of Sexuality in Colonial Andean Culture*. Austin, TX: University of Texas Press, 2005.

Howe, K. R. *Vaka Moana: Voyages of the Ancestors: The Discovery and Settlement of the Pacific*. Auckland: David Bateman, 2006.

Howell, Frederick W. W. *Icelandic Pictures Drawn with Pen and Pencil*. London: The Religious Tract Society, 1893.

Hunt, Terry. "Rethinking Easter Island's Ecological Catastrophe." *Journal of Archaeological Science* 34 (2007): 485–502.

Hunt, Terry, and Carl Lipo. *The Statues That Walked: Unraveling the Mystery of Easter Island*. Edited by Ethan E. Cochrane and Terry L. Hunt. New York: Free Press (Simon and Schuster), 2011.

Hunt, Terry, and Carl Lipo. "The Archaeology of Rapa Nui (Easter Island)." In *The Oxford Handbook of Prehistoric Oceania*, edited by Ethan E. Cochrane and Terry L. Hunt, 416–49. New York: Oxford University Press, 2018.

Hunter, Fraser, Martin Goldberg, Julia Farley, and Ian Leins. "In Search of the Celts." In *Celts: Art and Identity*, edited by Julia Farley and Fraser Hunter, 18–35. London: British Museum Press: National Museums, Scotland, 2015.

Ioannidis, A. G., et al. "Native American Gene Flow into Polynesia Predating Easter Island Settlement." *Nature* 583 (2020), 572–77.

Irwin, Alan. *Citizen Science: A Study of People, Expertise and Sustainable Development*. London: Routledge, 1995.

Irwin, Geoffrey. *The Prehistory Exploration and Colonization of the Pacific*. Cambridge: Cambridge University Press, 1992.

Jack, Jordynn. *Autism and Gender: From Refrigerator Mothers to Computer Geeks*. Champaign: University of Illinois Press, 2014.

Jackson, M. *The Secret Lives of Glaciers*. Brattleboro, VA: Green Writers Press, 2019.

Jackson, Rachael C., and Dorothy W. DeLaune. "Decolonizing Community Writing with Community Listening: Story, Transrhetorical Resistance, and Indigenous Cultural Literacy Activism." *Community Literacy Journal* 13, no. 1 (Fall 2018): 37–54.

James, Peter Bai et al. "Traditional and Complementary Medicine Use among Ebola Survivors in Sierra Leone: A Qualitative Exploratory Study of the Perspectives of Healthcare Workers Providing Care to Ebola Survivors." *BMC Complementary Medicine and Therapies* 20, no. 1 (2020). https://doi.org/10.1186/s12906-020-02931-6.

James, Peter Bai, Jon Wardle, Amie Steel, and Jon Adams. "Pattern of Health Care Utilization and Traditional and Complementary Medicine Use among Ebola Survivors in Sierra Leone." *PLoS ONE* 14, no. 9 (2019). https://doi.org/10.1371/journal.pone.0223068.

Johnson, Ayana Elizabeth, and Katharine K. Wilkinson, eds. *All We Can Save: Truth, Courage, and Solutions for the Climate Crisis*. New York: One World, 2020.

Johnson, Jenell. *American Lobotomy: A Rhetorical History*. University of Michigan Press, 2014.

Johnson-Sheehan, Richard, and Paul Lynch. "Rhetoric of Myth, Magic, and Conversion: A Prolegomena to Ancient Irish Rhetoric." *Rhetoric Review* 26, no. 3 (2007): 233–52.

Jones, Terry, et al. *Polynesians in America Pre-Columbian Contacts with the New World*. Lanham, MD: Altamire Press, 2011.

Kartha, Sivan. "Discourses of the Global South." In *The Oxford Handbook of Climate Change and Society*, ed. John S. Dryzek, Richard B. Norgaard, and David Schlosberg, 504–19. Oxford: Oxford University Press, 2011.

Kawaharada, Dennis. "The Discovery and Settlement of Polynesia." *Hawaiian Voyaging Traditions*. Accessed October 17, 2021. http://archive.hokulea.com/ike/moolelo/discovery_and_settlement.html.

Kay, Lily. *The Molecular Vision of Life: Caltech, The Rockefeller Foundation, and the Rise of the New Biology*. New York: Oxford University Press, 1993.

Kay, Samuel, Bo Zhao, and Daniel Sui. "Can Social Media Clear the Air? A Case Study of the Air Pollution Problem in Chinese Cities." *The Professional Geographer* 67, no. 3 (2015): 351–63.

Kaya, Hassan O. and Yonah N. Seleti. "African indigenous knowledge systems and relevance of higher education in South Africa," *International Education Journal: Comparative Perspectives* 12, no. 1 (2014): 30–44.

Keller, Evelyn Fox. *The Century of the Gene*. Cambridge: Harvard University Press, 2000.

Keller, Evelyn Fox. *Reflections on Gender and Science*. New Haven: Yale University Press, 1985.

Kennedy, George A. *Aristotle on Rhetoric: A Theory of Civic Discourse: Translated with Introduction, Notes and Appendices*. Oxford: Oxford University Press, 2007.

Keränen, Lisa. *Scientific Characters: Rhetoric, Politics, and Trust in Breast Cancer Research*. Tuscaloosa: University of Alabama Press, 2010.

Kerins, S. "The Future of Homelands/Outstations." *Dialogue* 29, no. 1 (2010): 52–60.

Keyser, Paul. *The Oxford Handbook of Science and Medicine in the Classical World*. Oxford: Oxford University Press, 2018.

Kimmerer, Robin Wall. *Braiding Sweetgrass: Indigenous Wisdom, Scientific Knowledge and the Teachings of Plants*. Minneapolis, MN: Milkweed Editions, 2013.

Kimmerer, Robin Wall. "Weaving Traditional and Ecological Knowledge into Biological Education: A Call to Action." *BioScience* 52, no. 5 (2002): 432–38.

Kinder, Marsha. *Playing with Power in Movies, Television, and Video Games: From Muppet Babies to Teenage Mutant Ninja Turtles*. Berkeley: University of California Press, 1991.

King, Anthony. "Polymer Tied in Celtic Knots." *Chemistry World*, May 2013. https://www.chemistryworld.com/news/polymer-tied-in-celtic-knots/6205.article.

King, Lisa, Rose Gubele, and Joyce Rain Anderson. *Survivance, Sovereignty, and Story: Teaching American Indian Rhetorics*. Boulder: University Press of Colorado, 2015.

Kingsley, Patrick. "Turkey Drops Evolution from Curriculum, Angering Secularists." *New York Times*, April 23, 2017. https://www.nytimes.com/2017/06/23/world/europe/turkey-evolution-high-school-curriculum.html.

Kirch, Patrick. *The Evolution of Polynesian Chiefdoms*. Cambridge: Cambridge University Press, 1984.

Kirch, Patrick. *A Shark Going Inland is My Chief: The Island Civilization of Ancient Hawai'i*. Berkeley: University of California Press, 2012.

Klett, Mark, Byron Wolfe, and Rebecca Solnit. *Yosemite in Time: Ice Ages, Tree Clocks, Ghost Rivers*. San Antonio: Trinity University Press, 2008.

Koban, John. "Walleye Wars and Pedagogical Management: Cooperative Rhetorics of Responsibility in Response to Settler Colonialism." *Rhetoric Society Quarterly* 50, no. 5 (2020): 321-34. https://doi.org/10.1080/02773945.2020.1813324.

Kovach, Margaret. *Indigenous Methodologies: Characteristics, Conversations and Contexts*. Toronto, ON: University of Toronto Press, 2009.

Kuhn, Thomas S. "Metaphor in Science." In *Metaphor and Thought*, edited by Andrew Ortony, 533-42. London: Cambridge University Press, 1979.

Thomas Kuhn, *Structure of Scientific Revolutions*. Chicago: University of Chicago Press, 2012.

Lacy, Terry G. *Ring of Seasons: Iceland—Its Culture & History*. Ann Arbor, MI: University of Michigan Press, 2000.

Ladefoged, Thegn N., et al. "Soil Nutrient Analysis of Rapa Nui Gardening." *Archaeology in Oceania* 45 (2010): 80-85.

LaFollette, Hugh, and Niall Shanks. *Brute Science: Dilemmas of Animal Experimentation*. London: Routledge, 2020.

Laliberté, André, and Marc Lanteigne. "The Issue of Challenges to the Legitimacy of CCP Rule." In *The Chinese Party-State in the 21st Century: Adaptation and the Reinvention of Legitimacy*, ed. André Laliberté and Marc Lanteigne, 1-21. London and New York: Routledge: 2008.

Langdon, Robert. "Manioc, a Long Concealed Key to the Enigma of Easter Island." *The Geographical Journal* 154, no. 3 (1988): 324-336.

Langton, Marcia. *Burning Questions: Emerging Environmental Issues for Indigenous Peoples in Northern Australia*. Darwin, Australia: Centre for Indigenous Natural and Cultural Resource Management, Northern Territory University, 1998.

Larrue, Sébastien, et al. "Anthropogenic Vegetation Contributions to Polynesia's Social Heritage." *Economic Botany* 64, no. 4 (15 December 2010): 329-39.

Latour, Bruno. "Visualization and Cognition: Thinking with Eyes and Hands." In *Knowledge and Society: Studies in the Sociology of Culture Past and Present*, edited by Elizabeth Long and Henrika Kuklick, 6:1-40. Stamford, CT: JAI Press, 1986.

Latour, Bruno. *We Have Never Been Modern*. Cambridge, MA: Harvard University Press, 1993.

Latz, Peter Kenneth, and Jenny Green. *Bushfires & Bushtucker: Aboriginal Plant Use in Central Australia*. Alice Springs, Australia: IAD Press, 1995.

Latz, Peter Kenneth, and G. F. Griffin. "Changes in Aboriginal Land Management in Relation to Fire and Food Plants in Central Australia." Symposium on the Nutrition of Aborigines, Canberra, Australia, October 23, 1978.

"Lecture on Iceland." *Birmingham Daily Post*, March 8, 1892.

Leitner, Helga, and Eric Sheppard. "From Kampungs to Condos? Contested Accumulations through Displacement in Jakarta." *Environment and Planning A: Economy and Space* 50, no. 2 (2018): 437–56.

Levins, Richard. "How Cuba is Going Ecological." In *Biology Under the Influence: Dialectical Essays on Ecology, Agriculture, and Health,* edited by Richard Lewontin and Richard Levins, 343–64. New York: Monthly Review Press, 2005.

Levins, Richard. "Science and Progress: Seven Developmentalist Myths in Agriculture." In *Biology Under the Influence: Dialectical Essays on Ecology, Agriculture, and Health,* edited by Richard Lewontin and Richard Levins, 321–28. New York: Monthly Review Press, 2005.

Levins, Richard and Richard Lewontin. *The Dialectical Biologist.* Cambridge, MA: Harvard University Press, 1985.

Lewontin, Richard. *Biology as Ideology.* New York: HarperCollins, 1991.

Lewontin, Richard. "The Maturing of Capitalist Agriculture: Farmer as Proletarian." In *Biology Under the Influence: Dialectical Essays on Ecology, Agriculture, and Health,* edited by Richard Lewontin and Richard Levins, 329–42. New York: Monthly Review Press, 2007.

Lewontin, Richard, and Richard Levins. *Biology Under the Influence: Dialectical Essays on Ecology, Agriculture, and Health.* New York: Monthly Review Press, 2005.

Lewis, David. *We, the Navigators: The Ancient Art of Landfinding in the Pacific.* 2nd ed. Honolulu: University of Hawai'i Press, 1994.

Liller, William. "The Megalithic Astronomy of Easter Island: Orientations of Ahu and Moai." *Journal for the History of Astronomy,* Archaeoastronomy Supplement 20 (1989): s21–48.

Liller, William. *The Ancient Solar Observatories of Rapanui: The Archaeoastronomy of Easter Island.* Old Bridge, NJ: Cloud Mountain Press, 1993.

Liller, William, and Julio Duarte. "Easter Island's Solar Ranging Device, Ahu Huri A Urenga, and Vicinity," *Archaeoastronomy* 9 (1986): 39–51.

Lim, Stephen S., Theo Vos, Abraham D Flaxman, Goodarz Danaei, Kenji Shibuya, Heather Adair-Rohani, Mohammad A. AlMazroa, et al. "A Comparative Risk Assessment of Burden of Disease and Injury Attributable to 67 Risk Factors and Risk Factor Clusters in 21 Regions, 1990–2010: A Systematic Analysis for the Global Burden of Disease Study 2010." *The Lancet* 380, no. 9859 (2012): 2224–60.

Lindenfeld, Laura A., et al. "Creating a Place for Environmental Communication Research in Sustainability Science." *Environmental Communication* 6, no. 1 (2012): 23–43. https://doi.org/10.1080/17524032.2011.640702.

Loo, Seng. "Islam, Science and Science Education: Conflict or Concord?" *Studies in Science Education* 36 (01/01 2001): 45–77. https://doi.org/10.1080/03057260108560167.

Love, Charles. "The Easter Island Moai Roads: An Excavation Project to Investigate the Roads Along Which the Easter Islanders Moved Their Gigantic Ancestral Statues." Report. Rock Springs: Western Wyoming Community College, 2001.

Lowe, Donald. *The Body in Late Capitalist USA*. Durham, NC: Duke University Press, 1995.

Lu, Si-ming. "A Case Study of Risk Communication: The Beijing Smog: The Communication Battle between the Public and Government." *DEStech Transactions on Social Science, Education, and Human Science*, 2016. https://doi.org/10.12783/dtssehs/emass2016/6804.

Lynch, Paul. "'*Ego Patricius, peccator rusticissimus*': The Rhetoric of St. Patrick of Ireland." *Rhetoric Review* 27, no. 2 (March 25, 2008): 111–30. https://doi.org/10.1080/07350190801921735.

Lynch, Paul, and Nathaniel Rivers. *Thinking with Bruno Latour in Rhetoric and Composition*. Carbondale, IL: SIU Press, 2015.

Lyons, Scott R. "Rhetorical Sovereignty: What Do American Indians Want from Writing?" *College Composition and Communication* 51, no. 3 (2000): 447–68.

MacDonald, Marjorie. "Complexity Science in Brief." University of Victoria, August 2012. https://www.uvic.ca/research/groups/cphfri/projects/currentprojects/complexity/ (webpage discontinued).

MacFarlane, Robert. *Landmarks*. London: Hamish Hamilton, 2015.

MacLachlan, Malcolm. *Culture and Health. A Critical Perspective towards Global Health. BMJ : British Medical Journal*. New York: John Wiley and Sons, 2006.

Maclean, K., C. J. Robinson, and O. Costello, eds. *A National Framework to Report on the Benefits of Indigenous Cultural Fire Management*. Collingwood, Australia: CSIRO, 2018.

Mahboubi, Mohaddese. "Effectiveness of Myrtus Communis in the Treatment of Hemorrhoids." *Journal of Integrative Medicine* 15, no. 5 (2017): 351–58. https://doi.org/10.1016/s2095-4964(17)60340-6.

Makemson, Maud Worcester. "Hawaiian Astronomical Concepts," *American Anthropologist* 40 (July–September 1938): 370–83.

Manguvo, Angellar, and Benford Mafuvadze. "The Impact of Traditional and Religious Practices on the Spread of Ebola in West Africa: Time for a Strategic Shift." *The Pan African Medical Journal* 22 (2015). https://doi.org/10.11694/pamj.supp.2015.22.1.6190.

Mann, Daniel, et al. "Drought, Vegetation Change, and Human History on Rapa Nui (Isla de Pascua, Easter Island)." *Quaternary Research* 69 (2008): 16–28.

Mann, Daniel, et al. "Prehistoric Destruction of the Primeval Soils and Vegetation of Rapa Nui (Isla de Pascua, Easter Island)." In *Easter Island: Scientific Exploration into the World's Environmental Problems in Microcosm*, edited by John Loret and John T. Tanacredi, 133–53. New York: Kluwer Academic–Plenum, 2003.

Manning, Jennifer. "A Decolonial Feminist Ethnography: Empowerment, Ethics and Epistemology." In *Empowering Methodologies in Organisational and Social Research*, 39–54. India: Routledge, 2022.

Mao, LuMing. "Doing Comparative Rhetoric Responsibly." *Rhetoric Society Quarterly* 41, no. 1 (2011): 64–69.

Mao, LuMing, Bo Wang, Arabella Lyon, Susan C. Jarratt, C. Jan Swearingen, Susan Romano, Peter Simonson, Steven Mailloux, and Xing Lu. "Manifesting a Future for Comparative Rhetoric." *Rhetoric Review* 34, no. 3 (2015): 239–74.

Marmion, Doug, Kazuko Obata, and Jakelin Troy. *Community, Identity, Wellbeing: The Report of the Second National Indigenous Languages Survey*. Canberra: Australian Institute of Aboriginal and Torres Strait Islander Studies, 2014.

Martínez-Alier, J. *The Environmentalism of the Poor: A Study of Ecological Conflicts and Valuation*. Cheltenham, UK: Edward Elgar Publishing, 2003. https://books.google.com/books?id=4JIzg4PUotcC.

Martinsson-Wallin, Hélène. *Ahu: The Ceremonial Stone Structures of Easter Island*. Uppsala, Sweden: Societas Archaeologica Upsaliensis, 1994.

Mavhunga, Clapperton Chakanetsa. *The Mobile Workshop: The Tsetse Fly and African Knowledge Production*. Cambridge, MA: MIT Press, 2018.

Mays, Chris. "Writing Complexity, One Stability at a Time: Teaching Writing as a Complex System." *College Composition and Communication* 68, no. 3 (2017): 559–85.

Mays, Chris. "'You Can't Make This Stuff Up': Complexity, Facts, and Creative Nonfiction." *College English* 80, no. 4 (2018): 319–41.

Mazzocchi, Fulvio. "Western Science and Traditional Knowledge: Despite Their Variations, Different Forms of Knowledge Can Learn from Each Other." *EMBO Reports* 7, no. 5 (2006): 463–66. https://doi.org/10.1038/sj.embor.740069.

McConvell, Patrick, Piers Kelly, and Sebastien Lacrampe. *Skin, Kin and Clan: The Dynamics of Social Categories in Indigenous Australia*. Sydney: ANU Press, 2018.

McCoy, Patrick C. "Easter Island." In *The Prehistory of Polynesia*, edited by J. D. Jennings. Cambridge, MA: Harvard UP, 1979.

McKemey, Michelle B. et al. "Cross-Cultural Monitoring of a Cultural Keystone Species Informs Revival of Indigenous Burning of Country in South-Eastern Australia." *Human Ecology* 47, no. 6 (2019): 893–904.

Medin, Douglas L., and Megan Bang. *Who's Asking?: Native Science, Western Science, and Science Education*. Cambridge, MA: MIT Press, 2014.

Menzies, Karen. "Understanding the Australian Aboriginal Experience of Collective, Historical and Intergenerational Trauma." *International Social Work* 62, no. 6 (2019): 1522–34.

Merchant, Carolyn. *The Death of Nature: Women, Ecology, and the Scientific Revolution*. New York: HarperCollins, 1980.

Merton, Robert K. *The Sociology of Science: Theoretical and Empirical Investigations*. Chicago: University of Chicago Press, 1973.

Métraux, Alfred. *Ethnology of Easter Island*. Honolulu: Bishop Museum Press, 1971.
Mistry, Jayalaxshmi, and Andrea Berardi. "Bridging Indigenous and Scientific Knowledge." *Science* 352, no. 6291 (2016): 1274–75.
Moorcroft, Heather. "Paradigms, Paradoxes and a Propitious Niche: Conservation and Indigenous Social Justice Policy in Australia." *Local Environment* 21, no. 5 (2016): 591–614.
Moore, John W., and Conrad L. Stanitski. *Chemistry: The Molecular Science*. Boston, MA: Cengage Learning, 2014.
Moore, Kristen R., and Daniel P. Richards. *Posthuman Praxis in Technical Communication*. New York: Routledge, 2018.
Morphy, F. "Australia's Indigenous Protected Areas: Resistance, Articulation and Entanglement in the Context of Natural Resource Management." In *Entangled Territorialities: Negotiating Indigenous Lands in Australia and Canada*, edited by Françoise Dussart and Sylvie Poirier, 70–90. Toronto, ON: University of Toronto Press, 2017.
Morton, Steve, Mandy Martin, Kim Mahood, and John Carty, eds. *Desert Lake: Art, Science and Stories from Paruku*. Collingwood, Australia: CSIRO, 2013.
Müller, Birgit. "The Temptation of Nitrogen: Fao Guidance for Food Sovereignty in Nicaragua." Paper presented at Food Sovereignty: A Critical Dialogue, International Conference, Yale University, CT, September, 2013.
Mulloy, William T. "Contemplate the Navel of the World." *Americas* 26.4 (1974): 25–33.
Mulloy, William T. "A Solstice Oriented Ahu on Easter Island," *Archaeology and Physical Anthropology in Oceania* 10 (April 1975), 1–39.
Mulloy, William T. "Double Canoes on Easter Island?" *Archaeology and Physical Anthropology in Oceania* 10.3 (Oct 1975): 181–84.
Mulloy, William T., and Gonzalo Figueroa. *The A Kivi-Vai Teka Complex and its Relationship to Easter Island Architectural Prehistory*. Honolulu: Social Science Research Institute, University of Hawai'i at Manoa, 1978.
Muñoz-Rodríguez, Pablo, et al. "Reconciling Conflicting Phylogenies in the Origin of Sweet Potato and Dispersal to Polynesia." *Current Biology* 28 (2018): 1246–56.
Nakata, Martin N. *Disciplining the Savages, Savaging the Disciplines*. Sydney, Australia: Aboriginal Studies Press, 2007.
National University of Ireland, Galway. "Polymer Breakthrough Inspired by Trees and Ancient Celtic Knots." ScienceDaily, May 22, 2013. https://www.sciencedaily.com/releases/2013/05/130522085335.htm.
Neale, Timothy, Rodney Carter, Trent Nelson, and Mick Bourke. "Walking Together: A Decolonising Experiment in Bushfire Management on Dja Dja Wurrung Country." *Cultural Geographies* 26, no. 3 (2019): 341–59.
Neidl, Phoebe. "Why *All We Can Save* Will Make You Feel Hopeful About the Climate Crisis." *Rolling Stone*, September 21, 2020. https://www.rolling

stone.com/culture/culture-news/all-we-can-save-book-climate-ayana-johnson-katharine-wilkinson-1062310.

Newmark, Julianne. "The Formal Conventions of Colonial Medicine: Bureau of Indian Affairs' Agency Physicians' Reports, 1880–1910." *College Composition and Communication* 71, no. 4 (2020): 620-42.

Noble, David. "Corporate Roots of American Science." In *Science and Liberation*, edited by Rita Arditti, Pat Brennan, Steve Cavrak, 63–75. Boston: South End Press, 1980.

Ogilvie, Astrid E. J. "Local Knowledge and Travellers' Tales: A Selection of Climatic Observations in Iceland." *Developments in Quaternary Science* 5 (2005), 257–87.

Ólafsson, Eggert, and Bjarni Pálsson. *Travels in Iceland*. Barnard and Sultzer, 1805.

Olivier, Laurent. "Les Codes De Représentation Visuelle Dans L'art Celtique Ancien (Visual Representation Codes in Early Celtic Art)." In *Celtic Art in Europe: Making Connections*, edited by Christopher Gosden, Sarah Crawford, and Katharina Ulmschneider, translated by S. Crawford, 39–55. Oxford and Philadelphia: Oxbow Books, 2014.

Olman, Lynda, and Danielle DeVasto. "Hybrid Collectivity: Hacking Environmental Risk Visualization for the Anthropocene." *Communication Design Quarterly* 8, no. 4 (2020): 18–28.

Orímóògùnjé, Oládélé Caleb. "The Yorùbá Indigenous Psychotherapeutic Healing System: A Case Study of Oríkì." *International Journal of Humanities and Cultural Studies (IJHCS)* 2, no. 4 (2016): 856–65.

Orliac, Catherine. "The Woody Vegetation of Easter Island Between the Early 14th and the Mid-17th Centuries AD." In *Easter Island Archaeology: Research on Early Rapanui Culture*, edited by C. S. Ayres and W. Ayres, 211–20. Los Osos, CA: Easter Island Foundation, 2000.

Orliac, Catherine. "Ligneux et palmiers de l'île de Pâques du XIéme a XVIIéme siécle de notre ére." In *Archéologie en Océanie insulaire: Peuplement, sociétés et paysages*, edited by C. Orliac, 184–99. Paris: Editions Artcom, 2003.

Orliac, Catherine, and Michel Orliac. *Easter Island: The Mystery of the Stone Giants*. New York: Abrams, 1995.

Oslund, Karen. *Iceland Imagined: Nature, Culture, and Storytelling in the North Atlantic*. Seattle, WA: University of Washington, 2011.

Pan, Shiyi. "Online Poll Requesting CMEP to Set up as Soon as Possible Enforceable Standard on Monitoring Pm 2.5." Weibo, November 6, 2011. Accessed November 21, 2020. https://www.weibo.com/panshiyi.

Pan, Shiyi. "Paying Attention to Beijing." Weibo, October 22, 2011. Accessed November 21, 2020, https://www.weibo.com/panshiyi.

Pan, Shiyi. "Interview with Luwei Luqiu." Weibo, November 3, 2011. Accessed November 21, 2020. https://www.weibo.com/panshiyi.

Pálsson, Sveinn. *Draft of a Physical, Geographical, and Historical Description of Icelandic Ice Mountains on the Basis of a Journey to Them in 1792-1794*.

Translated and edited by Richard S. Williams, Jr. and Oddur Sigurðsson. Reykjavík: Icelandic Literary Society, 2004.

Parreñas, Juno Salazar. "From Decolonial Indigenous Knowledges to Vernacular Ideas in Southeast Asia." *History and Theory* 59, no. 3 (2020): 413–20.

Parry, Claire Munoz. "Ebola: How a People's Science Helped End an Epidemic." *International Affairs* 93, no. 2 (2017): 485–86. https://doi.org/10.1093/ia/iix043.

"Particulate Matter (Pm. 2.5): Implementation of the 1997 National Ambient Air Quality Standards (NAAQS)." Wikileaks, 2008, accessed November 21, 2020, https://wikileaks.org/wiki/CRS:_Particulate_Matter_(PM2.5):_Implementation_of_the_1997_National_Ambient_Air_Quality_Standards_(NAAQS),_November_26,_2008.

Pascoe, B. *Dark Emu: Black Seeds: Agriculture or Accident?* Broome: Magabala Books, 2014.

Pavel, Pavel. "Reconstruction of the Transport of Moai." *State and Perspectives of Scientific Research in Easter Island Culture* (1990): 141–44.

Perelman, Chaim, and Lucie Olbrechts-Tyteca. *The New Rhetoric*. Notre Dame, IN: Notre Dame University Press, 1971.

Pestun, Aleksandr Vitalyevich, and Rustem Chingizovich Valeyev. "Шествие каменных голиафов: гипотеза [Procession of Stone Goliaths: Hypothesis]." In *На суше и на море* [*On Land and Sea*], edited by Boris Borobyov, 412–30. Moscow: Mysl Publishing, 1988.

Pestun, Aleksandr Vitalyevich, and Rustem Chingizovich Valeyev. *Моаи острова Пасхи. Инженерные решения древних загадок* [*The Moai of Easter Island. Engineering Solutions to Ancient Mysteries*]. Saint Petersburg: Petersburg XXI Century, 2016.

Peters, Annette, Douglas W. Dockery, James E. Muller, and Murray A. Mittleman. "Increased Particulate Air Pollution and the Triggering of Myocardial Infarction." *Circulation* 103, no. 23 (2001): 2810–15.

Petchey, Owen L., Peter J. Morin, and Han Olff. "The Topology of Ecological Interaction Networks: The State of the Art." In *Community Ecology: Processes, Models, and Applications*, edited by Herman Verhoef and Peter J. Morin, 7–22. Oxford: Oxford University Press, 2010.

Pietrucci, Pamela, and Leah Ceccarelli. "Scientist Citizens: Rhetoric and Responsibility in L'Aquila." *Rhetoric and Public Affairs* 22, no. 1 (2019): 95–128.

Podmore, Zac. "Navajo Nation Has a Higher Coronavirus Testing Rate Than Utah and Most States." *The Salt Lake Tribune*, April 19, 2020. https://www.sltrib.com/news/2020/04/19/navajo-nation-has-higher/.

Ponzi, Frank. *Islands Howells/Howell's Iceland: 1890–1901*. Mosfellsbær, Iceland: Brennholt, 2004.

Porter, David. *Journal of a Cruise Made to the Pacific Ocean*. 2nd ed. 3 vols. New York: Wiley & Halsted, 1822.

Powell, Malea. "Rhetorics of Survivance: How American Indians Use Writing." *College Composition and Communication* 53, no. 3 (2002): 396–434.

Pratt, Mary Louise. "Arts of the Contact Zone." *Profession* (1991): 33-40.

Randall, John E., and Alfredo Cea, *Shore Fishes of Easter Island*. Honolulu: University of Hawai'i Press, 2011.

Rasekoala, Elizabeth, and Lindy Orthia. "Anti-Racist Science Communication Starts with Recognising Its Globally Diverse Historical Footprint." *Impact of Social Sciences Blog*, July 1, 2020. https://blogs.lse.ac.uk/impactofsocialsciences/2020/07/01/anti-racist-science-communication-starts-with-recognising-its-globally-diverse-historical-footprint/.

Read, Peter. *A Rape of the Soul So Profound: The Return of the Stolen Generation*. London: Routledge, 2020.

Rickert, Thomas. *Ambient Rhetoric: The Attunements of Rhetorical Being*. Pittsburgh, PA: University of Pittsburgh Press, 2013.

Rieppel, Lukas, Eugenia Lean, and William Deringer. "Introduction: The Entangled Histories of Science and Capitalism." *OSIRIS* 33 (2018): 1–24. http://doi.org/10.1086/699170.

Rist, Stephan, and Farid Dahdouh-Guebas. "Ethnosciences—a Step Towards the Integration of Scientific and Indigenous Forms of Knowledge in the Management of Natural Resources for the Future." *Environment, Development and Sustainability* 8, no. 4 (2006): 467–93.

Ritchie, Joy, and Kathleen Boardman, "Feminism in Composition: Inclusion, Metonymy, and Disruption," *College Composition and Communication* 50, no. 4 (1999): 585–606.

Ritter, David. "The Rejection of Terra Nullius in Mabo: A Critical Analysis." *Sydney Law Review* 18 (1996): 5.

Roberts, Siobhan. "Flattening the Coronavirus Curve: One Chart Explains Why Slowing the Spread of the Infection is Nearly as Important as Stopping It." *The New York Times*, March 27, 2020. https://www.nytimes.com/article/flatten-curve-coronavirus.html/.

Rodriguez, Amardo. "A New Rhetoric for a Decolonial World." *Postcolonial Studies* 20, no. 2 (2017): 176–86.

Rose, Deborah Bird. "Exploring an Aboriginal Land Ethic." *Meanjin* 47, no. 3 (1988): 378–87.

Routledge, Katherine. *The Mystery of Easter Island*. London: Hazel, Watson and Viney, 1919.

Rowland-Shea, Jenny, et al. "The Nature Gap: Confronting Racial and Economic Disparities in the Destruction and Protection of Nature in America." *Center for American Progress,* July 21, 2020. https://www.americanprogress.org/issues/green/reports/2020/07/21/487787/the-nature-gap.

Rowse, Tim. "How We Got a Native Title Act." *The Australian Quarterly* 65, no. 4 (1993): 110–32.

Rowse, Tim. "Review: *The Contest for Aboriginal Souls: European Missionary Agendas in Australia*." *Aboriginal History* 42 (2018): 195–98.

Rull, Valentí. "The Deforestation of Easter Island." *Biological Reviews* 95 (2020): 124–41.
Russell, Lynette, and Ian J. McNiven. "Monumental Colonialism: Megaliths and The Appropriation of Australia's Aboriginal Past." *Journal of Material Culture* 3, no. 3 (1998): 283–99.
Russell-Smith, J. "Studies in the Jungle: People, Fire and Monsoon Forest." In *Archaeological Research in Kakadu National Park*, edited by R. Jones, 241–67. Canberra: Australian National Parks and Wildlife Service, 1985.
Russell-Smith, J., et al. "Aboriginal Resource Utilisation and Fire Management Practice in Western Arnhem Land, Monsoonal Northern Australia: Notes for Prehistory, Lessons for the Future." *Human Ecology* 25, no. 2 (1997): 159–95. http://dx.doi.org/10.1023/A:1021970021670.
Russell-Smith, J., et al. "Improving Estimates of Savanna Burning Emissions for Greenhouse Accounting in Northern Australia: Limitations, Challenges, Applications." *International Journal of Wildland Fire* 18, no. 1 (2009): 1–18. https://doi.org/doi:10.1071/WF08009.
Russell-Smith, J., P. J. Whitehead, and P. Cooke. *Culture, Ecology and Economy of Fire Management in Northern Australian Savannas: Rekindling the Wurrk Tradition*. Collingwood, Australia: CSIRO Publishing, 2009.
Safier, L. Zakarin, A. Gumer, M. Kline, D. Egli, and M. V. Sauer. "Compensating Human Subjects Providing Oocytes for Stem Cell Research: 9-Year Experience and Outcomes." *Journal of Assisted Reproduction and Genetics* 35 (2018): 1219–25.
Sahagún, Fray Bernardino de. *Historia General De Las Cosas De Nueva España*. Vol. 2. Ciudad México: Alianza Editorial Mexicana, 1989.
Saliba, George. *Islamic Science and the Making of the European Renaissance*. Cambridge, MA: MIT Press, 2007.
Salmón, Enrique. "Kincentric Ecology: Indigenous Perceptions of the Human–Nature Relationship." *Ecological Applications* 10, no. 5 (2000): 1327–32.
Savage, Gerald, and Godwin Agboka. "Guest Editors' Introduction to Special Issue." *Professional Communication, Social Justice, and the Global South* (2016): 3.
Sæþórsdóttir, Anna Dóra, C. Michael Hall, and Jarkko Saarinen. "Making Wilderness: Tourism and the History of the Wilderness Idea in Iceland." *Polar Geography* 34, no. 4 (2011): 249–73.
Seegert, Natasha. "Play of Sniffication: Coyotes Sing in the Margins." *Philosophy & Rhetoric* 47, no. 2 (2014): 158–78. https://doi.org/10.5325/philrhet.47.2.0158.
Selin, Helaine. *Encyclopaedia of the History of Science, Technology, and Medicine in Non-Westen Cultures*. Berlin, Germany: Springer Science & Business Media, 2013.
Semali, Ladislaus M., and Joe L. Kincheloe. *What is Indigenous Knowledge?: Voices from the Academy*. London: Routledge, 2002.

"Shanghai Expects to Release Pm 2.5 Data Next Year." *Top News*, November 6, 2011. http://news.sina.com.cn/green/news/roll/2011-11-15/065523467091.shtml.

Sharp, Andrew, ed. *The Journal of Jacob Roggeveen*. Oxford: Clarendon Press, 1970.

Shepardson, Britton. *Explaining Spatial and Temporal Patterns of Energy Investment in the Prehistory Statuary of Rapa Nui (Easter Island)*. PhD diss., University of Hawai'I at Mānoa, 2006.

Shepherd, Nan. *The Living Mountain*. Edinburgh: Canongate, 2019.

Sherwell, Phillip. "$40bn to Save Jakarta: The Story of the Great Garuda." *The Guardian*, November 22, 2016. https://www.theguardian.com/cities/2016/nov/22/jakarta-great-garuda-seawall-sinking.

Shiva, Vandana. *Biopiracy: The Plunder of Nature and Knowledge*. Berkeley: North Atlantic Press, 2016.

Sigurðardóttir, Ragnhildur, et al. "Trolls, Water, Time, and Community: Resource Management in the Mývatn District of Northeast Iceland" In *Global Perspectives on Long Term Community Resource Management*, edited by Ludomir R. Lozny and Thomas H. McGovern, 77–101. Cham, Switzerland: Springer Nature Switzerland, 2019.

Sigurðsson, Oddur, and Richard S. Williams, Jr. *Geographic Names of Iceland's Glaciers: Historic and Modern*. Reston, VA: US Geological Survey, 2008.

Silver, Daniel S. "Knot Theory's Odd Origins: The Modern Study of Knots Grew out an Attempt by Three 19th-Century Scottish Physicists to Apply Knot Theory to Fundamental Questions about the Universe." *American Scientist* 94, no. 2 (2006): 158–65.

Sithole, B. et al. *Aboriginal Land and Sea Management in the Top End: a Community Driven Evaluation*. Darwin, Australia: CSIRO Publishing, 2007.

Sivasundaram, Sujit. "Sciences and the Global: On Methods, Questions, and Theory." *Isis* 101, no. 1 (2010): 146–58.

Skottsberg, Carl. *The Natural History of Juan Fernandez and Easter Island*. Vol. 1. Delhi: Alpha Editions, 2020. (Reprint of Uppsala: Almqvist & Wiksells Boktryckeri, 1956.)

Slaughter, James C., Thomas Lumley, Lianne Sheppard, Jane Q. Koenig, and Gail G. Shapiro. "Effects of Ambient Air Pollution on Symptom Severity and Medication Use in Children with Asthma." *Annals of Allergy, Asthma & Immunology* 91, no. 4 (2003): 346–53.

Smith, Kevin P. "*Landnám*: The Settlement of Iceland in Archaeological and Historical Perspective." *World Archaeology* 26, no. 3 (1995): 319–47.

Smith, Linda Tuhiwai. *Decolonizing Methodologies: Research and Indigenous Peoples*. London: Zed Books, 1999.

Smyth, D., P. Taylor, and A. Willis, eds. *Aboriginal Ranger Training and Employment in Australia, Proceedings of the First National Workshop*. Canberra: Australian National Parks and Wildlife Service, 1985.

Soetan, Olusegun. "Charms and Amulets." In *Culture and Customs of the Yoruba*, edited by Toyin Falola and Akintunde Akinyemi, 205–13. Austin: Pan-African University Press, 2017.
South, Eugenia C., et al. "Effect of Greening Vacant Land on Mental Health of Community-Dwelling Adults: A Cluster Randomized Trial." *Journal of the American Medical Association* 1, no. 3 (2018): e180298. https://doi.org/10.1001/jamanetworkopen.2018.0298.
South, Eugenia C., Michelle C. Kondo, Rose A. Cheney, Charles C. Branas. "Neighborhood Blight, Stress, and Health: A Walking Trial of Urban Greening and Ambulatory Heart Rate" *American Journal of Public Heath* 105, no. 5 (2015): 909–13. https://doi.org/10.2105/AJPH.2014.302526.
Steffensen, Victor. *Fire Country: How Indigenous Fire Management Could Help Save Australia*. Collingwood, Australia: CSIRO Publishing, 2020.
Stevens, Stan. "Indigenous Management." In *Conservation through Cultural Survival: Indigenous Peoples and Protected Areas*, edited by S. Stevens, 189–224. Washington: Island Press, 1997.
Stevenson, Christopher M., et al. "Variation in Rapa Nui Land Use Indicates Production and Population Peaks Prior to European Contact." *Proceedings of the National Academy of Sciences*, 112, no. 4 (2015): 1025–30.
Stokes, Leah Cardamore. "A Field Guide for Transformation." In *All We Can Save: Truth, Courage, and Solutions for the Climate Crisis*, edited by Ayana Elizabeth Johnson and Katharine K. Wilkinson, 337–47. New York: One World, 2020.
Stone, John. "Fifty Years of Unremitting Failure: Aboriginal Policy Since the 1967 Referendum." *Quadrant* 61, no. 11 (2017): 62–72.
Storey, Alice, et al. "Radiocarbon and DNA Evidence for a Pre-Columbian Introduction of Polynesian Chickens to Chile." *Proceedings of the National Academy of Sciences* 104, no. 25 (June 19, 2007): 10335–39.
Stroud, Scott R. "Pragmatism and the Methodology of Comparative Rhetoric." *Rhetoric Society Quarterly* 39, no. 4 (2009): 353–79.
Sturtevant, William C. "Studies in Ethnoscience 1." *American Anthropologist* 66, no. 3 (1964): 99–131.
Sunder Rajan, Kaushik. *Biocapital: The Constitution of Postgenomic Life*. Durham, NC: Duke University Press, 2006.
Sunder Rajan, Kaushik. "Introduction," in *Lively Capital*, edited by Kaushik Sunder Rajan. Durham, NC: Duke University Press, 2012.
Susskind, Lawrence. "Complexity Science and Collaborative Decision Making." *Negotiation Journal* 26, no. 3 (July 1, 2010): 367–70. https://doi.org/10.1111/j.1571-9979.2010.00278.x.
Swift. Jonathan. "A Modest Proposal for Preventing the Children of Poor People in Ireland from Being a Burden on Their Parents or Country and for Making

Them Beneficial to the Publick," Project Gutenberg. https://www.gutenberg.org/files/1080/1080-h/1080-h.htm.

TallBear, Kim. "Indigenous Bioscientists Constitute Knowledge across Cultures of Expertise and Tradition: An Indigenous Standpoint Research Project." In *Re:Mindings: Co-Constituting Indigenous, Academic, Artistic Knowledges*, edited by May-Britt Öhman, Hiroshi Maruyama, Johan Gärdebo, 173–91. Uppsala, Finland: The Hugo Valentin Centre, 2014. https://urn.kb.se/resolve?urn=urn:nbn:se:uu:diva-383415.

Thioune, Oumar, Sidy Dieng, Ahmedou Bamba Koueimel Fall, and Moussa Diop. "Contribution of Nanotechnology in the Improvement of the Anti-Inflammatory Activity of Shea Butter." *American Journal of PharmTech Research* 9, no. 6 (2019): 242–53. https://doi.org/10.46624/ajptr.2019.v9.i6.021.

Thomson, Vicki, et al. "Using Ancient DNA to Study the Origins and Dispersal of Ancestral Polynesian Chickens Across the Pacific." *Proceedings of the National Academy of Sciences* 111, no. 13 (April 1, 2014): 4826–31.

Þorgilsson, Ari. *Landnámabók*. Reykjavík: Sigurður Kristjánsson, 1891.

Thorsby, Erik, et al. "Further Evidence of an Amerindian Contribution to the Polynesian Gene Pool on Easter Island." *Tissue Antigens* 73, no. 6 (2009): 582–85.

Thorsby, Erik, et al. "The Polynesian Gene Pool: an Early Contribution by Amerindians to Easter Island." *Philosophical Transactions: Biological Sciences* 367, no. 1590 (March 19, 2012): 812–19.

Tofail, Syed A. M., Elias P. Koumoulos, Amit Bandyopadhyay, Susmita Bose, Lisa O'Donoghue, and Costas Charitidis. "Additive Manufacturing: Scientific and Technological Challenges, Market Uptake and Opportunities." *Materials Today* 21, no. 1 (January 1, 2018): 22–37. https://doi.org/10.1016/j.mattod.2017.07.001.

Tonino, Leah, "Two Ways of Knowing: Robin Wall Kimmerer on Scientific and Native American Views of the Natural World," *Sun Magazine*, April 2016. https://www.thesunmagazine.org/issues/484/two-ways-of-knowing.

Tuck, Eve, and K. Wayne Yang. "Decolonization Is Not a Metaphor." *Tabula Rasa* 38 (2021): 61–111.

Van Tilburg, JoAnne. *Easter Island: Archaeology, Ecology and Culture*. London: British Museum Press, 1994.

Van Tilburg, JoAnne. *Remote Possibilities: Hoa Hakananai'a and HMS Topaze on Rapa Nui*. British Museum Research Papers, 158. London: British Museum Press, 2006.

Verma, Nandini, Rina Chakrabarti, Rakha H. Das, and Hemant K. Gautam. "Anti-Inflammatory Effects of Shea Butter through Inhibition of Inos, COX-2, and Cytokines via the NF-KB Pathway in Lps-Activated J774 Macrophage Cells." *Journal of Complementary and Integrative Medicine* 9, no. 1 (2012): 1–11. https://doi.org/10.1515/1553-3840.1574.

Vigfússon, Guðbrandur. *Bárðarsaga Snæfellsáss, Viglundarsaga: Þorðarsaga, Draumavitranir, Völsaþáttr.* Copenhagen: Det Nordiske Literatur-Samfund, 1860.

Vigilante, T. "Analysis of Explorers' Records of Aboriginal Landscape Burning in the Kimberley Region of Western Australia." *Australian Geographical Studies* 39, no. 2 (2001): 135–55.

Vigilante, T. "The Ethnoecology of Landscape Burning Around Kalumburu Aboriginal Community, North Kimberley Region, Western Australia." PhD diss., Charles Darwin University, 2004.

Vincent, Warwick, F. "Arctic Climate Change: Local Impacts, Global Consequences, and Policy Implications." In *The Palgrave Handbook of Arctic Policy and Politics*, edited by Ken S. Coates and Carin Holroyd, 507–27. London: Palgrave Macmillan, 2020.

Vora, Kalindi. *Life Support: Biocapital and the New History of Outsourced Labor.* Minneapolis: University of Minnesota Press, 2015.

Wainwright, Joel. *Decolonizing Development: Colonial Power and the Maya.* New York: John Wiley & Sons, 2011.

Waldby, Catherine, and Melinda Cooper. "The Biopolitics of Reproduction: Post-Fordist Biotechnology and Women's Clinical Labor." *Australian Feminist Studies* 23, no. 55 (2008): 57–73. https://doi.org/10.1080/08164640701816223.

Waldby, Catherine, and Melinda Cooper. "From Reproductive Work to Regenerative Labour: The Female Body and the Stem Cell Industries." *Feminist Theory* 11, no. 3 (2010): 3–22. https://doi.org/10.1177/1464700109355210.

Waldby, Catherine, Ian Kerridge, Margaret Boulos, and Katherine Carroll. "From Altruism to Monetisation: Australian Women's Ideas About Money, Ethics, and Research Eggs." *Social Science & Medicine* 94 (2013): 34–42. https://doi.org/10.1016/j.socscimed.2013.05.034.

Waldrop, M. Mitchell. "Viewing the Universe as a Coat of Chain Mail." *Science* 250 (December 14, 1990): 1510–11.

Walker, Kenneth C. *Climate Politics on the Border: Environmental Justice Rhetorics.* Tuscaloosa: University of Alabama Press, 2022.

Walker, Mark. "Pandemic Highlights Deep-Rooted Problems in Indian Health Service." *New York Times*, October 1, 2020, 8. https://www.nytimes.com/2020/09/29/us/politics/coronavirus-indian-health-service.html.

Wallaby, Catherine. *The Oocyte Economy: The Changing Meaning of Human Eggs.* Durham, NC: Duke University Press, 2019.

Walsh, Lynda. "The Visual Rhetoric of Climate Change." *WIREs Climate Change* 6, no. 4 (2015): 361–68. https://doi.org/10.1002/wcc.342.

Walsh, Lynda, and Casey Boyle. *Topologies as Techniques for a Post-Critical Rhetoric.* New York: Springer, 2017.

Walsh, Lynda, and Kenneth C. Walker. "Perspectives on Uncertainty for Technical Communication Scholars." *Technical Communication Quarterly* 25, no. 2 (2016): 71–86.

Wang, Bo. "Comparative Rhetoric, Postcolonial Studies, and Transnational Feminisms: A Geopolitical Approach." *Rhetoric Society Quarterly* 43, no. 3 (2013): 226–42.

Wang, Tao. "Joining by 6 ENGOs, Southern Weekly Submitted Seven Suggestions to CMEP on the Second Draft of Ambient Air Quality Standard (for Suggestions)." *Southern Weekly*, November 24, 2011. http://www.infzm.com/content/65151.

Wanzer-Serrano, Darrel. "Decolonial Rhetoric and a Future yet-to-Become: A Loving Response." *Advances in the History of Rhetoric* 21, no. 3 (2018): 326–30.

White, Hayden. *Metahistory: The Historical Imagination in Nineteenth-Century Europe*. Baltimore, MD: Johns Hopkins University Press, 2014 (1973).

Wilder, Kelley. *Photography and Science*. Chicago: University of Chicago, 2009.

Williams, Miriam F., "Reimagining NASA: A Cultural and Visual Analysis of the U.S. Space Program." *Journal of Business and Technical Communication* 26, no. 3 (2012): 368–89.

Wilson, Shawn. *Research is Ceremony: Indigenous Research Methods*. Halifax, NS: Fernwood Publishing, 2008.

Winkel, Eric. "Tawhw and Science: Essays on the History and Philosophy of Islamic Science, by Osman Bakar (Review)." *Muslim World* LXXXIII, no. 3-4 (1993): 329–35.

Wiseman, Nathanael D., and Douglas K. Bardsley. "Climate Change and Indigenous Natural Resource Management: A Review of Socio-Ecological Interactions in the Alinytjara Wilurara NRM region." *Local Environment* 18, no. 9 (2013): 1024–45.

Wong, K. Chimin, and Lien-teh Wu. *History of Chinese Medicine: Being a Chronicle of Medical Happenings in China from Ancient Times to the Present Period*. Tientsin, China: Tientsin Press, 1932.

Wood, Ellen Meiksins. *Democracy Against Capitalism: Renewing Historical Materialism*. London: Verso, 2016.

Wynn, James. *Citizen Science in the Digital Age: Rhetoric, Science, and Public Engagement*. Tuscaloosa: University of Alabama Press, 2017.

Yibarbuk, D., et al. "Fire Ecology and Aboriginal Land Management in Central Arnhem Land, Northern Australia: A Tradition Of Ecosystem Management." *Journal of Biogeography* 28, no. 3 (2001): 325–43. http://dx.doi.org/10.1046/j.1365-2699.2001.00555.x.

Young, Elspeth A. 1992. "Aboriginal Land Rights in Australia: Expectations, Achievements and Implications." *Applied Geography* 12, no. 2 (1992): 146–61. https://doi.org/https://doi.org/10.1016/0143-6228(92)90004-7.

Young, Michael J. L., John Derek Latham, and Robert Bertram Serjeant, eds. *Religion, Learning and Science in the 'Abbasid Period*. London: Cambridge University Press, 2006.

Yoxen, Edward. "Life as a Productive Force: Capitalising the Science and Technology of Molecular Biology." In *Science, Technology, and the Labour Process: Marxist Studies Volume 1*, edited by Les Levidow and Bob Young, 66–122. London: CSE Books, 1981.

Zhang, Yanshuang. "Microblogging and Its Implications to Chinese Civil Society and the Urban Public Sphere: A Case Study of Sina Weibo." PhD diss., University of Queensland, 2015.

Zheng, Yuanjie. "Online Survey for People's Opinion on Beijing's Air Quality." Weibo, 2011. Accessed November 21, 2020. https://www.weibo.com/zhyj.

Zhao, Tianyu, Yu Zheng, Julien Poly, and Wenxin Wang. "Controlled Multi-Vinyl Monomer Homopolymerization Through Vinyl Oligomer Combination as a Universal Approach to Hyperbranched Architectures." *Nature Communications* 4, no. 1 (May 21, 2013): 1–8. https://doi.org/10.1038/ncomms2887.

Contributor Biographies

Joseph Bartolotta, Assistant Professor in the Department of Writing Studies & Rhetoric, Hofstra University, examines the training and application of usability and user experience principles in writing programs and for students in TPC. He further explores the ways schools and industry organizations define best practices, competencies, and ethics in their respective contexts, and looks for ways to bring both together for generative discussions.

Bridget Campbell is a PhD candidate in the School of Natural Sciences at Macquarie University. Passionate about cross-cultural ecology, Campbell is working closely with Yolŋu Yirralka Rangers, northeast Arnhem Land, for her PhD. She aspires to bridge gaps between Yolŋu and Euro-American scientific knowledge systems to investigate and ultimately safeguard biocultural diversity in the Laynhapuy Indigenous Protected Area.

Renee Cawthorne is a proud Wiradjuri woman who grew up on Darkinjung Country on the Central Coast of New South Wales, Australia. Renee graduated from Macquarie University with a Bachelor of Science in 2015 and is currently enrolled in a Master of Research in the School of Natural Science, Macquarie University, reviewing the pedagogies and processes in Indigenous Science units across Australian universities.

Jianfen Chen is an assistant professor in technical communication at Towson University. Her research interests include intercultural technical and professional communication, environmental risk communication, public health risk communication, and digital rhetoric. Her work appears in the proceedings of *ACM (Association of Computing Machinery) International*

Conference on Design of Communication and *International Journal of Humanities and Social Science*.

Sunnie R. Clahchischiligi is a member of the Navajo Nation and has a PhD in rhetoric and writing from the University of New Mexico (UNM). She earned her master's degree in rhetoric, writing, and digital media studies from Northern Arizona University. She is a 2022–2023 Russell J. and Dorothy S. Bilinski Fellow and received UNM's Outstanding Core Writing Instructor award for 2019–2020. Clahchischiligi is a longtime, award-winning journalist whose work appears in *Searchlight New Mexico, The New York Times, Rolling Stone* magazine, *The Guardian, The Navajo Times*, and many other publications.

Patrick Cooke is a Ganggalida man and master's student in the Cross-Cultural Ecology and Environmental Management Lab, School of Natural Sciences, Macquarie University, Sydney, Australia. He is conducting research on the Aboriginal dispersal of culturally significant Bunya Pine (*Araucaria bidwillii*) across the east coast of Australia.

Huiling Ding, Professor of English, is Director of the Master of Science in Technical Communication program at North Carolina State University. She is the author of *Rhetoric of a Global Epidemic: Transcultural Communication about SARS*, which received the National Council of Teachers of English (NCTE) Conference on College Composition and Communication (CCCC) 2016 Best Book Award in Technical and Scientific Communication. In addition, her articles won the 2013 Nell Ann Pickett Award for Best Article in *Technical Communication Quarterly* and the 2008 Editor's Pick New Scholar Award from *Written Communication*. Ding's research focuses on intercultural professional communication, health communication, risk communication, rhetoric of health and medicine, social justice, digital rhetoric, workplace communication, scientific communication, and comparative rhetoric. She serves on the editorial boards of leading journals such as *Technical Communication Quarterly, Communication Design Quarterly, Written Communication,* and *Rhetoric, Globalization, and Professional Communication*.

Evelyn Dsouza recently completed her PhD in rhetoric and scientific and technical communication at the University of Minnesota, Twin Cities. Her writing has appeared in *Programmatic Perspectives* and *Open Rivers:*

Rethinking Water, Place & Community. Today she works as a technical editor for the heritage and cultural resources division of a large civil engineering and environmental consulting firm. She enjoys teaching, reading, and ongoing interdisciplinary learning.

Ryan Eichberger is a Visiting Assistant Professor in English and the Writing Program at St. Olaf College, Minnesota. He is broadly interested in more-than-human and animal rhetoric, especially where these interests cross paths with visual and technical communication. His current teaching focuses on community-engaged environmental learning and place-based knowledge, while his current research focuses on ecological mourning and species conservation in the Great Lakes region. When he is not researching or teaching, he is probably out photographing birds or working in the garden.

Emilie Ens has conducted cross-cultural ecological research since 2008 and leads Macquarie University's Cross-Cultural Ecology and Environmental Management Lab. With her Indigenous colleagues she has published many papers on cross-cultural and multidisciplinary research and land management methods and received several awards for outstanding collaborations, including the 2017 Eureka Prize for Innovation in Citizen Science.

Monica Fahey is a PhD student in the School of Natural Sciences, Macquarie University, and the Research Centre for Ecosystem Resilience, Royal Botanic Gardens Sydney. Fahey has a background in the molecular ecology of native Australian plants and cultural anthropology. She is currently working on her PhD thesis, "Retracing the Dispersal of Rainforest Food Trees by Precolonial Aboriginal People," using plant genomic data.

Kelly Happe is an Associate Professor of Women's Studies and Communication Studies at the University of Georgia. She is the author of the award-winning book *The Material Gene: Gender, Race, and Heredity After the Human Genome Project* and co-editor of *Biocitizenship: The Politics of Bodies, Governance, and Power*. In 2019 she was awarded the University of Georgia Creative Research Medal in the Humanities and Arts.

Francisco Nahoe holds a PhD in Renaissance Literature from the University of Nevada and a ThM in Biblical Studies from Harvard Divinity School. As both an ethnic Rapa Nui and scion of American archaeologists, he would say that each side of his family has endowed him with complex and multifaceted views of the island and its culture. A Roman Catholic priest and Franciscan friar, he teaches rhetoric and philosophy at Zaytuna College in Berkeley, California.

Julianne Newmark is the Assistant Chair for Core Writing and Director of the Technical and Professional Communication program at the University of New Mexico. Dr. Newmark's publications have considered the school-to-work transition, multimodal community creation in online classrooms, and usability/UX/UCD. She also teaches, conducts research, and publishes in Indigenous studies, particularly concerning early 20th-century Native activist writers' rhetorically impactful bureaucratic writing, especially in Bureau of Indian Affairs contexts. She is currently at work on her second book, *Reports of Agency: Retrieving Indigenous Professional Communication in Dawes Era Indian Bureau Documents*, and continues to serve as editor-in-chief of *Xchanges*, a writing studies e-journal.

Lynda C. Olman is a Professor of English at the University of Nevada, Reno. She studies the rhetoric of science—particularly the public reception of visual arguments and of the ethos or public role of the scientist. Her most recent monograph, *Scientists as Prophets: A Rhetorical Genealogy* (Lynda Walsh, Oxford University Press, 2013), traces a dominant strand in the ethos of late-modern science advisers back to its historical roots in religious rhetoric. She is the recipient of a 2022 Friedrich Wilhelm Bessel Research Award to support work contributing to a forthcoming book (co-authored with Birgit Schneider) on the global visualization of forests for climate management.

Toluwani Oloke is an Assistant Professor of Public Relations and Health Communication at the University of North Georgia, Gainesville. Her scholarly focus is on the importance of cultural sensitivity in developing effective health campaign messages targeted at African audiences. She holds a PhD in Public Relations from the University of Florida and has taught an array of courses in international pr and health communication.

Shaina Russell is a postdoctoral research fellow in the School of Natural Sciences at Macquarie University. Russell is a cross-cultural ecologist passionate about bringing Indigenous knowledge and Euro-American science together for biocultural conservation. She is currently working with

Yolŋu in the Laynhapuy Indigenous Protected Area in northeast Arnhem Land on cross-cultural biodiversity research. Russell's PhD research focused on cross-cultural knowledge for enhanced understanding of freshwater Country in southeast Arnhem Land, northern Australia.

Sabina Rysnik-Steck recently completed her Master of Research in the Department of Earth and Environmental Sciences at Macquarie University. Rysnik-Steck works across Eora and Darug lands and employs cross-cultural collaborative practices. Her interdisciplinary thesis brought together GIS mapping techniques and Traditional Indigenous Knowledge to determine the impact of sea-level rise on Indigenous cultural sites around Sydney Harbour and Indigenous management preferences.

Olusegun Soetan is a film specialist and a cultural studies scholar with a focus on the Nigerian film industry and African indigenous knowledge systems. He holds a PhD in African languages and literature from the University of Wisconsin–Madison. He is an Assistant Teaching Professor of African Studies at the Pennsylvania State University where he teaches and researches on African cinema, indigenous science/knowledge systems, and African popular culture and languages. He has published essays in peer-reviewed journals and contributed book chapters to edited volumes. He is a poet, novelist, and photographer.

Daniel Sloane is a PhD student in the Cross-Cultural Ecology and Environmental Management Lab, School of Natural Sciences, Macquarie University, Sydney, Australia. He works with Yolŋu in the Laynhapuy Indigenous Protected Area to combine Indigenous knowledge and Euro-American science to better understand and manage climatic and biological impacts on eco-culturally significant forests.

Index

African sciences, 63–83
African traditional cultures, 63–83
Ahu, 120, 123, 125–131
air quality monitoring, 87–89, 92–109
arctic, 179–80, 192–193
art history, 139–40, 142–145

biocapital 10, 24, 32–33, 36–37

capital, 23–38
capitalism, 23–38
Celtic, 139–152
China, 87–109
climate change, 2, 8, 11, 13, 41, 45, 117, 140, 180, 183, 191–192, 194–195
colonization, x, 3, 11, 13, 29, 41, 45–47, 49, 54, 82, 129. *See also* rhetorics, colonial
coproduction (of science and capital), 24, 26, 30, 37
COVID-19, 14–15, 23, 70, 157–158, 160–166, 168, 174–176
commensality, 183, 196
complexity, 14, 90, 140, 145–146, 148–151
community health communication, 63–83, 87–109
critical contextualized method, 87–89, 92–93, 106, 109

decolonization, 3, 7, 9, 11, 14–15, 16, 49, 56, 69
Diné (see also, Navajo Nation), 14–15, 157–176

Easter Island. *See* Rapa Nui
Ebola, 11–2, 63–65, 69–70, 76–80, 82
ecocide, 130
ecological grief, 181, 183–184, 192, 194–196
environmental communication, 87–109, 179–197
environmental management 2, 11, 45–57
epidemics. *See* COVID-19; Ebola
epistemology, ix–xi, 25, 30, 65, 69, 141
ethnomathematics, ix–x
Euro-American science:
    definition of, 1n
    global dominance of, 5–6, 9–10, 23–38
    interaction with global sciences, xi, 1–2, 5–6, 10–12, 14–16, 50, 76–82, 107–108, 184–191
expropriation, 24, 26–30, 33–35, 37

fire management 54–56

glaciers, 179–197

Index

health sciences, 63–83, 87–109, 157–176
home ground, 181–184
hoʻokele, 116
hybrid forums, 89–90, 97, 106–109

Iceland, 179–197
Indigenous Australians, 1–2, 45–57
Indigenous kinship systems, 46, 50–51, 53
Indigenous land rights, 46, 48–49
Indigenous Protected Area, 49, 51–52, 57
Indigenous Rangers, 52, 55
irony, 10–12, 45–57, 63–83, 87–109

labor, 23–38, 123–132
life, capitalization of, 26–38

mana, 119, 125–126, 131
marae, 118, 123, 131
matamuʻa, 114–115, 122, 124–128, 130–132
mathematics, ix–xi, 3, 141, 145–147
metaphor, ix, 12–14, 113–132, 139–152
metonymy, 14–15, 157–176, 179–197
Micronesia, 113–132
Moai, 113–132

Native Title, 48–49
Navajo Nation, 14–15, 157–176. *See also* Diné

oocytes, 33–34, 42

PM 2.5, 87–89, 92–109. *See also* air quality monitoring

Polynesian wayfinding, 113, 116–118, 125, 131

Rapa Nui, 113–132
rephotography, 189
rhetoric(s):
 classical, 4, 6–7
 colonial, 47–49. *See also* colonization
 comparative, 7
 decolonial, 7, 49, 56, 69
 definition, 3–4, 7
 Indigenous, ix–xi, 7, 45–57, 63–83, 113–132, 139–152, 157–176, 179–197
 Islamic/Classical Arabic, 4–6
 of science, 3, 7–9
 visual, 113–132, 139–152, 157–176
rhetorical sequencing, 141
risk, 3–4, 6–8, 32, 88–98, 105, 107, 109, 170–171

stem cells, 24, 33–37
synecdoche 9–10, 23–38

transcultural risk communication, 90–92
tropes, 9, 130–131, 160. *See also* irony; metaphor; synecdoche; metonymy

uncertainty, 6, 90–91, 132

value, 24, 26, 29–37, 52–54, 64, 90, 103–107, 141–144, 166–168, 173–175, 182

Yoruba, 68–76, 82

www.ingramcontent.com/pod-product-compliance
Lightning Source LLC
Chambersburg PA
CBHW030538230426
43665CB00010B/942